Student Solutions Manual
Volume 1

to accompany

Physics for Engineers and Scientists

THIRD EDITION

Student Solutions Manual
Volume 1

to accompany

Physics for Engineers and Scientists

THIRD EDITION

Hans Ohanian / John Markert

by

Stephen Luzader
Hang Deng-Luzader
David Marx

W. W. NORTON AND COMPANY
NEW YORK   LONDON

ISBN 978-0-393-92979-9

W. W. Norton & Company, Inc., 500 Fifth Avenue, New York, NY 10110
www.wwnorton.com
W. W. Norton & Company Ltd., Castle House, 75/76 Wells Street, London W1T 3QT

1 2 3 4 5 6 7 8 9 0

CONTENTS

PREFACE TO STUDENT SOLUTIONS MANUAL

The *Student Solutions Manual* contains complete solutions to approximately half the problems at the end of each chapter in *Physics for Engineers and Scientists* by Hans Ohanian and John Markert. These solutions include more detailed explanations than those in the separate *Instructor Solutions Manual*.

This manual should be used in a variety of ways. If a problem has been assigned for which the solution is given in this manual, then you should make every effort to solve it before consulting the solution given here. Additionally, this manual gives detailed solutions to problems that may not be assigned by your instructor that will assist you in solving similar problems. You should examine closely the thought processes used by the authors and emulate them in your own problem solving.

Generally, the final numerical answers have been given to the same number of significant figures as the data in the problems, unless intermediate answers are needed for subsequent parts of the problem. In such cases, numerical results may be presented with more significant figures than the problem data. However, slight discrepancies in the last significant figure may occur as a result of individual preferences in rounding and use of different calculators from those used in preparation of the manual. Every effort has been made to ensure the accuracy of the solutions. If students or instructors discover errors in this manual, the authors would greatly appreciate being notified so corrections can be issued.

We would like to acknowledge all those who have helped in the development and production of this manual:

Lowell Boone, University of Evansville
Jason Stevens, Deerfield Academy
Ray Zich, Illinois State University
Stiliana Antonova, Barnard College
Krassi Lazarova, Ph.D., Drexel University
Rebecca Grossman, Columbia University
Brian Arneson, Sapling Systems
William T. Younger III, College of the Albemarle

Steven Luzader, Frostburg State University
Hang Deng-Luzader, Frostburg State University
David Marx, Illinois State University

CHAPTER 1 SPACE, TIME, AND MASS

1-7.   Let's convert 1/2 inch to mm using conversion factors, then use proportional reasoning to do the others until the number of significant figures becomes large.

$$\frac{1}{2} \text{ in} \times 25.4 \text{ mm/in} = 12.7 \text{ mm}$$

$$\frac{1}{4} \text{ in} = \frac{12.7 \text{ mm}}{2} = 6.35 \text{ mm}$$

$$\frac{1}{8} \text{ in} = \frac{6.35 \text{ mm}}{2} = 3.175 \text{ mm} = 3.18 \text{ mm} \text{ (to three significant figures)}$$

$$\frac{1}{16} \text{ in} = \frac{3.175 \text{ mm}}{2} = 1.5875 \text{ mm} = 1.59 \text{ mm} \text{ (to three significant figures)}$$

The number of digits is becoming large, so let's do direct conversions for the rest of the problems.

$$\frac{1}{32} \text{ in} \times 25.4 \text{ mm/in} = 0.794 \text{ mm}$$

$$\frac{1}{64} \text{ in} \times 25.4 \text{ mm/in} = 0.397 \text{ mm}$$

1-13.   1 turn = 360°, so 5° × 1 turn/360° = 0.0139 turn. For an English thread,

$$0.0139 \text{ turn} \times \frac{1 \text{ in}}{80 \text{ turns}} \times \frac{0.0254 \text{ m}}{\text{in}} \times \frac{10^6 \text{ }\mu\text{m}}{\text{m}} = 4.41 \text{ }\mu\text{m}. \text{ For a metric thread,}$$

$$0.0139 \text{ turn} \times \frac{0.5 \text{ mm}}{\text{turn}} \times \frac{10^{-3} \text{ m}}{\text{mm}} \times \frac{10^6 \text{ }\mu\text{m}}{\text{m}} = 6.94 \text{ }\mu\text{m}.$$

1-15.   For one of the triangles, $(R + 1.75 \text{ m})^2 = R^2 + (4700 \text{ m})^2$. Expand this to get $R^2 + 2(1.75 \text{ m})R + (1.75 \text{ m})^2 = R^2 + (4700 \text{ m})^2$.

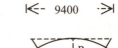

We expect R to be much larger than 1.75 m, so we can ignore $(1.75 \text{ m})^2$ relative to all the other terms. The R^2 terms cancel, leaving
$(3.50 \text{ m})R = (4700 \text{ m})^2$, which gives $R = \underline{6.3 \times 10^6 \text{ m}}$.

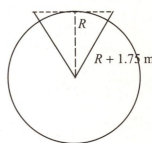

1-21. $1 \text{ sidereal day} \times \dfrac{365.25 \text{ solar days/year}}{366.25 \text{ sidereal days/year}} \times \dfrac{24 \text{ h}}{\text{solar day}} = 23.934 \text{ h/sidereal day}$. Using 1 h = 60

min to convert the 0.934 h to minutes gives 1 sidereal day = 23 h 56 min.

1-27.   1 day = 24 hr = 86,400 s. The earth rotates 360° per day, which corresponds to a rotation rate of
$\dfrac{360°}{86,400 \text{ s}} \times \dfrac{60 \text{ min}}{\text{degree}} = 0.250 \text{ min/s}$. A timing error of 1 s will result in an angular error of 0.250 min. According to Problem 1-14, 1 min = 1852 m, so the corresponding error in position is 0.250 min/s × 1852 m = 463 m = 0.463 km.

1

1-29. $m_{planets} = (0.33 + 4.9 + 5.98 + 0.64 + 1900 + 553 + 87.3 + 10.8 + 0.66) \times 10^{24}$ kg $= 2.56 \times 10^{27}$ kg

(to three significant figures). $m_{sun} = 1.99 \times 10^{30}$ kg, so the total mass is

$m_{total} = m_{sun} + m_{planets} = 1.99 \times 10^{30}$ kg $+ 2.56 \times 10^{27}$ kg $= 1.99 \times 10^{30}$ kg (to three significant

figures). The fraction of the total mass included in the planets is

$\dfrac{m_{planets}}{m_{total}} \times 100\% = \dfrac{2.56 \times 10^{27}}{1.99 \times 10^{30}} \times 100\% = 0.134\%$. The fraction of the mass in the sun is $\underline{100\% -}$

$\underline{0.134\% = 99.9\%}$.

1-31.   From the periodic table in the Appendix, we see that the uranium nucleus contains about 238

nucleons, each with the mass of a proton. Table 1.7 gives 1.7×10^{-27} kg for the mass of a proton.

Then the total mass of the electrons is $(92)(9.1 \times 10^{-31}$ kg$) = 8.4 \times 10^{-29}$ kg, and the total mass of

the nucleus is $(238)(1.7 \times 10^{-27}$ kg$) = 4.0 \times 10^{-25}$ kg. To two significant figures, the total mass of

the atom is 4.0×10^{-25} kg. The fraction of the total mass in the electrons is $8.4 \times 10^{-29}/4.0 \times 10^{-25}$

$= 2.1 \times 10^{-4} = \underline{0.021\%}$. The fraction of mass in the nucleus is $\underline{99.98\%}$.

1-37.   Molecular mass of $N_2 = 28$ g/mol

Molecular mass of $O_2 = 32$ g/mol

Molecular mass of Ar $= 40$ g/mol

Therefore, 1000 g of air will contain:

755 g $N_2 = 755$ g/(28 g/mol) $= 27.0$ mol

232 g $O_2 = (232/32)$ mol $= 7.25$ mol

13 g Ar $= (13/40)$ mol $= 0.325$ mol

The percentage by number of molecules of these substances is:

N_2: $27.0/(27.0 + 7.25 + 0.325) \times 100\% = 27.0/34.575 \times 100\% = 78.1\%$

O_2: $(7.25/34.575) \times 100\% = 21\%$

Ar: $(0.325/34.575) \times 100\% = 0.9\%$

Therefore, the "molecular mass" of air is

$(0.781 \times 28) + (0.21 \times 32) + (0.009 \times 40) = \underline{28.95\,\text{g/mol}}$.

1-39. $0.53° = 0.53/360 \times 2\pi$ rad $= 9.25 \times 10^{-3}$ rad

$d = 9.25 \times 10^{-3}$ rad $\times 1.5 \times 10^{11}$ m

$\quad = 1.4 \times 10^{9}$ m

$r = 6.9 \times 10^{8}$ m

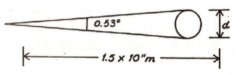

1-45. $1\,\text{m}^3 \times \left(\dfrac{1\ \text{ft}}{0.3048\ \text{m}}\right)^3 = \underline{35.31\ \text{ft}^3}$. Note that the conversion factor must be cubed.

1-47.   His height is measured to a precision of 0.1 inch, so we want to see how many significant digits

this implies. 8 feet $= 96$ inches (exactly), so to the nearest 0.1 inch his height can be expressed as

107.1 inches, which contains four significant figures. Converting 11.1 inches to feet gives his

height in feet: 8 ft $+ 11.1$ in $= 8.925$ ft, which also contains four significant figures. Thus his

height in meters should be specified to four figures: 8.925 ft $\times 0.3048\ \dfrac{\text{m}}{\text{ft}} = \underline{2.720\ \text{m}}$.

1-49. $8.9 \text{ g/cm}^3 \times \left(\dfrac{1 \text{ kg}}{1000 \text{ g}} \right) \times \left(\dfrac{100 \text{ cm}}{\text{m}} \right)^3 = 8.9 \times 10^3 \text{ kg/m}^3$. Note that the cm to m conversion factor

must be cubed! 1 ft = 0.3048 m, 1 lb = 0.454 kg.

$8.9 \times 10^3 \text{ kg/m}^3 \times \left(\dfrac{1 \text{ lb}}{0.454 \text{ kg}} \right) \times \left(\dfrac{0.3048 \text{ m}}{\text{ft}} \right)^3 = 555 \text{ lb/ft}^3$, or $\underline{5.6 \times 10^2 \text{ lb/ft}^3}$ to two significant

figures. Again, note that the m to ft conversion factor must be cubed. 1 ft = 12 in, so

$555 \text{ lb/ft}^3 \times \left(\dfrac{1 \text{ ft}}{12 \text{ in}} \right)^3 = 0.32 \text{ lb/in}^3$.

1-53. $N = \dfrac{1 \text{ cm}^2 \times \dfrac{10^{-4} \text{ m}^2}{\text{cm}^2}}{10^{-12} \dfrac{\text{m}^2}{\text{transistor}}} = \underline{10^8 \text{ transistors}}$. If they're stacked, N = the number of transistors per

layer (10^8) × number of layers. If the cube is 1 cm high and each layer is 10^{-7} m, or 10^{-5} cm, thick, then the cube holds 10^5 layers and the cube can hold $10^8 \times 10^5 = \underline{10^{13} \text{ transistors}}$!

1-55. (a) $(3.6 \times 10^4) \times (2.049 \times 10^{-2}) = (3.6)(2.049) \times 10^{4-2} = \underline{7.4 \times 10^2}$
(b) $(2.581 \times 10^2) - (7.264 \times 10^1) = (2.581 - 0.7264) \times 10^2 = \underline{1.855 \times 10^2}$

(c) $0.079832 \div 9.43 = \dfrac{7.9832 \times 10^{-2}}{9.43 \times 10^0} = 0.847 \times 10^{-2-0} = \underline{8.47 \times 10^{-3}}$

1-57. $\text{Density} = \dfrac{m}{V} = \dfrac{m}{\dfrac{4}{3} \pi R^3} = \dfrac{3m}{4 \pi R^3} = \dfrac{3(2.0 \times 10^{30} \text{ kg})}{4 \pi (20 \times 10^3 \text{ m})^3} \times \dfrac{1 \text{ metric ton}}{10^3 \text{ kg}} \times 10^{-6} \dfrac{\text{m}^3}{\text{cm}^3}$

$= \underline{6.0 \times 10^7 \text{ metric tons/cm}^3}$

1-59. From Table 1.10, 1 liter = 10^{-3} m³. According to data given in the "Conversion of Units" section of the chapter, the density of water is 1000 kg/m³, so

$300 \text{ liters/min} \times \dfrac{10^{-3} \text{ m}^3}{\text{liter}} \times \dfrac{1 \text{ min}}{60 \text{ s}} = 5.00 \times 10^{-3} \text{ m}^3/\text{s}$, and

$5.00 \times 10^{-3} \text{ m}^3/\text{s} \times 1000 \text{ kg/m}^3 = 5.00 \text{ kg/s}$.

1-61. This can be solved using proportional reasoning. $\text{Density} = \dfrac{m}{V} = \dfrac{m}{\dfrac{4}{3} \pi R^3} = \dfrac{3m}{4 \pi R^3}$, which means

$\dfrac{m_{lead}}{R_{lead}^3} = \dfrac{m_{copper}}{R_{copper}^3}$, from which we get $R_{lead} = R_{copper} \left(\dfrac{m_{lead}}{m_{copper}} \right)^{1/3} = (4.8 \times 10^{-15} \text{ m}) \left(\dfrac{3.5}{1.06} \right)^{1/3}$

$= \underline{7.1 \times 10^{-15} \text{ m}}$. Likewise we get $R_{oxygen} = R_{copper} \left(\dfrac{m_{oxygen}}{m_{copper}} \right)^{1/3} = (4.8 \times 10^{-15} \text{ m}) \left(\dfrac{0.27}{1.06} \right)^{1/3}$

$= \underline{3.0 \times 10^{-15} \text{ m}}$. Note that it was not necessary to include the factor ($\times 10^{-25}$) when expressing the masses because it cancels in the ratio.

1-65.   The slope is the tangent of the required angle. For a slope of 1:5, $\theta = \tan^{-1}\dfrac{1}{5} = \underline{11°}$. For a slope of

1:10, $\theta = \tan^{-1}\dfrac{1}{10} = \underline{5.7°}$. For a slope angle of 0.1°, $\tan\theta = 1.7 \times 10^{-3}$, so the slope is

1:$(1.7 \times 10^{-3})^{-1} = 1: 5.7 \times 10^2$, so the rise is about 1 atom per 570 atoms.

1-67.   In this calculation, assume that the calendar year is exactly 365 days long and a circle contains exactly 360°, so these are not interpreted as numbers with only three significant figures. The

angle the earth moves through in one calendar year is $\theta = \dfrac{365}{365.24} \times 360° = \underline{359.76°}$. In four

years, including one leap year with exactly 366 days, the total number of days is $3 \times 365 + 366 = $

1461 days. The angle the earth moves through in four years is $\theta = \dfrac{1461}{365.24} \times 360° = \underline{1440.0°}$,

which is exactly four complete circles to five significant figures. (The angle is actually a little larger than 1440.0°, which is why there are some four-year intervals that do not include an extra day.)

1-71. $m_{atom} = \dfrac{M}{N_A} = \dfrac{235\text{ g/mol}}{6.02204 \times 10^{23}\,\dfrac{\text{atoms}}{\text{mol}}} = 3.902 \times 10^{-22}\text{ g} = \underline{3.902 \times 10^{-25}\text{ kg}}$. In atomic mass units,

this is $m_{atom} = \dfrac{3.902 \times 10^{-25}\text{ kg}}{1.66054 \times 10^{-27}\,\dfrac{\text{kg}}{\text{u}}} = \underline{235.0\text{ u}}$.

1-83.   The arc AB has a length of 3900 km, so the half angle θ is 1950 km/R_E, where R_E is the radius of the earth. Using the value from Table 1-1 gives $\theta = (1950\text{ km})/(6.4 \times 10^3\text{ km}) = 0.305$ radian $= 17.5°$.
(a) The linear distance d from A to B is
$d = 2R_E \sin\theta = 2(6.4 \times 10^3\text{ km})(\sin 17.5°) = \underline{3840}$
$\underline{\text{km}}$.
(b) The depth h at the midpoint is
$h = R_E - R_E \cos\theta = R_E(1 - \cos\theta)$, which gives
$h = (6.4 \times 10^3\text{ km})(1 - \cos 17.5°) = \underline{296\text{ km}}$.

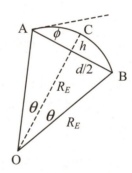

(c) "Horizontal" means parallel to the surface of the earth, or tangent to the surface of the earth at that location. The tangent line is shown in the diagram, and the slope angle ϕ between the tangent and the line AB is shown. Since the radius OA is perpendicular to the tangent line and the radius OC is perpendicular to the line AB, the angle ϕ must be the same as θ. Thus $\phi = 17.5°$ and the slope is the tangent of that angle: $\underline{\text{slope} = \tan 17.5° = 0.315}$. Using the ratio form of slope gives $\underline{\text{slope} = 1:(0.315)^{-1} = 1:3.2}$. (This is a quite steep slope!) Note that to observers standing at each end, the tunnel appears to be sloping down into the earth.

2-3.   1 year $= 3.156 \times 10^7$ sec, so 20 m/year $= \dfrac{20 \text{ m}}{1 \text{ year} \times 3.156 \times 10^7 \text{ s/year}} = 6.34 \times 10^{-7}$ m/s (6.3 ×

10^{-7} m/s to two significant figures). 1 day = 24 hr = 86,400 s. In cm/day the rate is
6.34×10^{-7} m/s $\times 8.64 \times 10^4$ s/day $\times 100$ cm/m = 5.4 cm/day.

2-11. $t = \dfrac{d}{v} \Rightarrow t_1 = \dfrac{1 \text{ km}}{4 \text{ cm/year}} = \dfrac{1000 \text{ m}}{4 \times 10^{-2} \text{ m/year}} = 2.5 \times 10^4$ years. Moving 1000 km will take

1000 times as long, or $t_2 = 2.5 \times 10^7$ years.

2-15.   Use the formula: $t = \dfrac{d}{v}$. $t_{air} = \dfrac{5280 \text{ km}}{900 \text{ km/hr}} = 5.87$ hr. $t_{shiip} = \dfrac{5280 \text{ km}}{35 \text{ km/hour}} = 151$ hr.

2-19.   (a) Take $x = 0$ to be the cheetah's starting position. Then the cheetah's position is given by
$x_c = v_c t$. The antelope's starting position is 50 m from the cheetah's starting position, so the
position of the antelope is given by $x_a = v_a t + 50$. When the cheetah catches the antelope, their
positions are the same, and we get $v_c t = v_a t + 50$. The speeds are $v_c = 101$ km/h = 28.1 m/s and

$v_a = 88$ km/h = 24.4 m/s. Solving the equation for t gives $t = \dfrac{50 \text{ m}}{v_c - v_a} = \dfrac{50 \text{ m}}{28.1 \text{ m/s} - 24.4 \text{ m/s}} =$

13.8 s, or 14 s to two significant figures. During this time, the cheetah travels (28.1 m/s)(13.8 s)
= 380 m.
(b) The cheetah must catch the antelope within 20 s. Call the antelope's initial position x_0.
We use the same equation that says the cheetah catches the antelope, $v_c t = v_a t + x_0$, but now we
set $t = 20$ s and calculate what head start x_0 the antelope needs. We get
$x_0 = (v_c - v_a)t = (28.1 \text{ m/s} - 24.4 \text{ m/s})(20 \text{ s}) = 72$ m. If the antelope is farther away than 72 m,
the cheetah will not be able to catch it.

2-21. $d = 26$ mi $\times 1.6 \times 10^3$ m/mi $+ 385$ yd $\times 0.9144$ m/yd $= 4.195 \times 10^4$ m
$t = 2$ hr 24 min 52 s $= 2$ hr $\times 3600$ s/hr $+ 24$ min $\times 60$ s/min $+ 52$ s $= 8692$ s

average speed $= \dfrac{d}{t} = \dfrac{4.195 \times 10^4 \text{ m}}{8692 \text{ s}} = 4.83$ m/s

2-23. $x = 4.0t - 0.50t^2$. To find the maximum value of x, differentiate with respect to t and set the

derivative equal to zero: $\dfrac{dx}{dt} = 4.0 - t = 0$. The result is $t = 4.0$ s. (This is the point at which the

runner turns around and moves back toward the starting line.) The distance traveled at this time is
$x = 4.0(4.0) - 0.50(4.0)^2 = 8.0$ m. At $t = 8$ seconds, $x = 4.0(8.0) - 0.5(8.0)^2 = 0$; that is when he
comes back to the starting line. The total distance traveled is 16 m. Then average speed $=$

$\dfrac{\text{distance}}{\text{time}} = \dfrac{16 \text{ m}}{8 \text{ s}} = 2.0$ m/s.

2-25.

Planet	Orbit circumference (km)	Period (s)	Speed (km/s)	log speed	log radius
Mercury	3.64×10^8	7.61×10^6	47.8	1.68	8.56
Venus	6.79×10^8	1.94×10^7	35.0	1.54	8.83
Earth	9.42×10^8	3.16×10^7	29.8	1.47	8.97
Mars	1.43×10^9	5.93×10^7	24.1	1.38	9.16
Jupiter	4.89×10^9	3.76×10^8	13.0	1.11	9.69
Saturn	8.98×10^9	9.31×10^8	9.65	0.985	9.95
Uranus	1.80×10^{10}	2.65×10^9	6.79	0.832	10.26
Neptune	2.83×10^{10}	5.21×10^9	5.43	0.735	10.45
Pluto	3.71×10^{10}	7.83×10^9	4.74	0.676	10.57

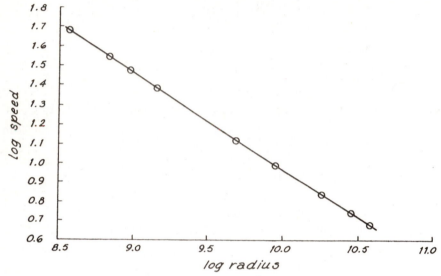

The slope of the line through the nine points is $-\dfrac{1}{2}$.

This means that log speed $= -\dfrac{1}{2}$ log radius + log C.

Therefore log speed $= \log[C(\text{radius})^{-1/2}]$. Thus
Speed $= C(\text{radius})^{-1/2}$ where C is some constant.

2-29.   Distance = (12 blocks + 6 blocks + 3 blocks) × 81 m/block = 1701 m.

The elapsed time = 14 min 5 s + 6 min 28 s + 3 min 40 s = 23 min 73 s = 1453 s. Then average

speed $= \dfrac{\text{distance}}{\text{time}} = \dfrac{1701 \ \text{m}}{1453 \ \text{s}} = 1.17$ m/s. The total displacement is $\Delta x =$ (12 blocks − 6 blocks +

3 blocks) × 81 m/block = 729 m. The average velocity is $\bar{v} = \dfrac{\Delta x}{\Delta t} = \dfrac{729 \ \text{m}}{1453 \ \text{s}} = 0.502$ m/s.

2-31.   Total distance = 3 × 0.25 mile × 1609 m/mile = 1207 m. Displacement = 0 m because the horse

returns to the starting point. Total time = 1 min 40 s = 100 s. Then average speed =

$\dfrac{\text{distance}}{\text{time}} = \dfrac{1207 \ \text{m}}{100 \ \text{s}} = 12.1$ m/s. The average velocity is $\bar{v} = \dfrac{\Delta x}{\Delta t} = 0.$

2-33. From the graph, the total distance traveled by the squirrel is 6 m + 6 m + 2 m + 6 m = 20 m. The total elapsed time is 30 s. Then average speed $= \dfrac{\text{distance}}{\text{time}} = \dfrac{20 \text{ m}}{30 \text{ s}} = \underline{0.67 \text{ m/s}}$. The total displacement is $\Delta x = 16$ m, so the average velocity is $\bar{v} = \dfrac{\Delta x}{\Delta t} = \dfrac{13 \text{ m}}{8.1 \text{ s}} = \underline{0.53 \text{ m/s}}$.

2-37. $\bar{a} = \dfrac{v_2 - v_1}{t_2 - t_1} = \dfrac{27 \text{ m/s} - 0}{8.0 \times 10^{-3} \text{ s}} = \underline{3.4 \times 10^3 \text{ m/s}^2}$.

2-41. $x = 3.6t^2 - 2.4t^3$, $v = \dfrac{dx}{dt} = 7.2t - 7.2t^2$. $v = 0$ if

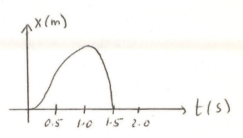

$7.2t - 7.2t^2 = 0 \Rightarrow t = 0$ s or $t = 1.0$ s. At $t = 0, x = 0$.
At $t = 1.0$ s, $x = 3.6 - 2.4 = 1.2$ m. To make a sketch, consider that $x = 0$ when $t = 0$ and $t = \dfrac{3.6}{2.4} = 1.5$ s.
Also, $dx/dt = 0$ when $t = 0$ and $t = 1$ s.

2-43. For $0 \le t \le 5.0$ s, $\bar{a} = \dfrac{v(t = 5 \text{ s}) - v(t = 0)}{5 \text{ s}} \approx \dfrac{5 \text{ m/s} - 0}{5 \text{ s}} = \underline{1 \text{ m/s}^2}$. (Your value may be slightly different depending on how you read the values of v at 0 and 5 s.) For $5.0 \le t \le 10.0$ s,

$\bar{a} = \dfrac{v(t = 10 \text{ s}) - v(t = 5 \text{ s})}{5 \text{ s}} \approx \dfrac{9.5 \text{ m/s} - 5.0 \text{ m/s}}{5 \text{ s}} = \underline{0.9 \text{ m/s}^2}$. (Again, your value may be slightly different depending on how you estimate the values of v at 5 and 10 s.) To find the instantaneous acceleration at 3 s, draw a tangent to the curve at that time. Your estimate may be slightly different from ours. We get a tangent line that passes through the points (1 s, 0 m/s) and (5 s, 5 m/s). The slope of this line is the instantaneous acceleration $a = \dfrac{5 \text{ m/s} - 0}{5 \text{ s} - 1 \text{ s}} = \underline{1.3 \text{ m/s}^2}$.

2-45. $a = \dfrac{dv}{dt} = \dfrac{d}{dt}\left[\dfrac{v_0}{(1 + At^2)}\right] = -\dfrac{v_0}{(1 + At^2)^2}\dfrac{d}{dt}(At^2) = -\dfrac{2Av_0t}{(1 + At^2)^2}$. $\underline{\text{At } t = 0, a = 0}$.

at $t = 2$ s, $a = -\dfrac{2(25 \text{ m/s})(2 \text{ s}^{-2})(2 \text{ s})}{\left[1 + (2 \text{ s}^{-2})(2 \text{ s})^2\right]^2} = \underline{-2.5 \text{ m/s}^2}$. As $t \to \infty$, $a \to -\dfrac{2Av_0t}{A^2t^4}$, so $a \to 0$.

2-47. (a)

(b)

Time interval (s)	Avg speed (m/s)	Distance traveled (m)
0–0.3	647.5	194
0.3–0.6	628.5	189
0.6–0.9	611.5	183
0.9–1.2	596.0	179
1.2–1.5	579.5	174
1.5–1.8	564.0	169
1.8–2.1	549.5	165
2.1–2.4	535.0	161
2.4–2.7	521.0	156
2.7–3.0	508.0	152

Total distance traveled = 1722 m

(c) Counting the number of squares under the v versus t curve gives the same answer within about ± 2 m.

2-49.   (a)

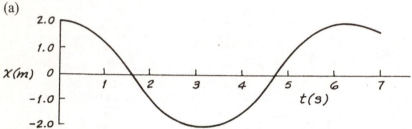

(b) Calculus method:

If $x = 0$, then $\cos t = 0$, so $t = \pi/2$ or $3\pi/2$ s. Particle crosses $x = 0$ at $t = 1.6$ s and 4.7 s.

$v = dx/dt = -2.0 \sin t$

$v(\pi/2 \text{ s}) = -2.0 \sin(\pi/2) = -2.0$ m/s

$v(3\pi/2 \text{ s}) = -2.0 \sin(3\pi/s) = 2.0$ m/s

$a = dv/dt = -2.0 \cos t$

$a(\pi/2) = -2.0 \cos(\pi/2) = 0 \text{ m/s}^2$

$a(3\pi/2) = -2.0 \cos(3\pi/2) = 0 \text{ m/s}^2$

(c) Maximum distance achieved when $\cos t = \pm 1$, i.e., when $t = 0$, π, 2π, or $t = 0$ s, 3.1s and 6.3s.

$v = -2.0 \sin t$

$v(0) = -2.0 \sin(0) = 0 \text{ m/s}$

$v(\pi) = -2.0 \sin(\pi) = 0 \text{ m/s}$

$v(2\pi) = -2.0 \sin(2\pi) = 0 \text{ m/s}$

$a = dv/dt = -2.0 \cos t$

$a(0) = -2.0 \cos(0) = -2.0 \text{ m/s}^2$

$a(\pi) = -2.0 \cos(\pi) = 2.0 \text{ m/s}^2$

$a(2\pi) = -2.0 \cos(2\pi) = -2.0 \text{ m/s}^2$

2-57. Use $\frac{1}{2}(v^2 - v_0^2) = a(x - x_0)$ with $x - x_0 = 50$ m; $v = 0$;

$v_0 = 96$ km/h $= 26.67$ m/s

$$a = \frac{\frac{1}{2}(v^2 - v_0^2)}{x - x_0} = \frac{\frac{1}{2}(-26.67)^2}{50 \, \text{m/s}^2}$$

$\underline{a = -7.1 \, \text{m/s}^2}$

Use $v = v_0 + at$ with $v = 0$ m/s, $v_0 = 96$ km/h $= 26.7$ m/s, and $a = -7.1$ m/s². Then

$t = -\frac{v_0}{a} = \underline{3.8 \, \text{s}}.$

2-63. The sketch should be based on the following:

For $0 \leq t \leq 6$ s, $a = 3.0$ m/s²; $v = v_0 + at = 0 + 3t$; $x = \int_0^t v \, dt = \int_0^t 3t \, dt = \frac{3}{2}t^2$. At $t = 6$ s,

$v = 3 \times 6 = \underline{18 \text{ m/s}}$; $x = \frac{3}{2} \cdot 6^2 = \underline{54 \text{ m}}$. For $6 \leq t \leq 10$ s, $a = -4.5 \text{ m/s}^2$, and

$v = v_0 + a(t - 6) = 18 - 4.5 \quad (t - 6) \text{ m/s} = 45 - 4.5t \text{ m/s}.$

$x = x_0 + v_0(t - 6) + \frac{1}{2}a(t - 6)^2 = 54 + 18(t - 6) - \frac{1}{2} \cdot 4.5(t - 6)^2 = -2.25(t - 6)^2 + 18(t - 6) + 54$

At $t = 10$ s, $v = 45 - 4.5 \times 10 = \underline{0}$; $x = -2.25 \cdot 4^2 + 18 \cdot 4 + 54 = \underline{90 \text{ m}}.$

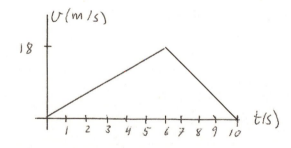

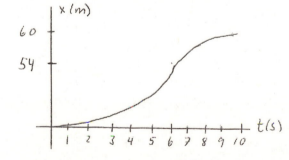

2-65. $x = v_0 t + \frac{1}{2}at^2 \Rightarrow v_0 = \dfrac{x - \dfrac{at^2}{2}}{t} = \dfrac{550 \, \text{m} - 0.5 \times 0.5 \, \text{m/s}^2 \times 15^2 \, \text{s}^2}{15 \, \text{s}} = 32.9 \, \text{m/s}.$

$v = v_0 + at = 32.9 \text{ m/s} + 0.5 \text{ m/s}^2 \cdot 15 \text{ s} = \underline{40.4 \text{ m/s}}.$

2-73. $v = -g\tau + g\tau e^{-t/\tau}$

(a) acceleration, $\dfrac{dv}{dt} = \underline{ge^{-t/\tau}}$

(b) $\underset{t \to \infty}{\mathrm{Lim}}\, e^{-t/\tau} = 0$, therefore

$\underset{t \to \infty}{\mathrm{Lim}}\, v = -g\tau + \underset{t \to \infty}{\mathrm{Lim}}\, e^{-t/\tau} = \underline{-g\tau}$

(c) $v = \dfrac{d}{dt}\left(-g\tau t - g t^2 e^{-t/\tau} + g\tau^2 + x_0\right)$

$= -g\tau + g\,\dfrac{\tau^2}{\tau}\, e^{-t/\tau} = \underline{-g\tau + g\tau e^{-t/\tau}}$

(d) for $t \ll \tau$, the exponential $e^{-t/\tau}$ can be expanded as

$e^{-t/\tau} = 1 - \dfrac{t}{\tau} + \dfrac{1}{2}\dfrac{t^2}{\tau^2} + \ldots\ldots$

Therefore, $x = -g\tau\, t - g\tau^2(1 - \dfrac{t}{\tau} + \dfrac{1}{2}\dfrac{t^2}{\tau^2}) + g\tau^2 + x_0$

$= -g\tau\, t - g\tau^2 + g t\,\tau - \dfrac{1}{2} g t^2 + g\tau^2 + x_0 = \underline{-\dfrac{1}{2} g t^2 + x_0}$

2-75.   130 km/h = 36.1 m/s. The acceleration of the freely falling falcon is g, so use

$v^2 - v_0^2 = 2g(x - x_0) \Rightarrow x - x_0 = \dfrac{v^2 - v_0^2}{2g} = \dfrac{(36.1 \text{ m/s})^2 - 0}{2(9.81 \text{ m/s}^2)} = 66.5 \text{ m}.$

2-83. $t_{up} = \sqrt{\dfrac{2h}{g}} = \sqrt{\dfrac{2(9.5 \text{ m})}{9.81 \text{ m/s}^2}} = 1.39$ s. Total time = 2(1.39 s) = $\underline{2.8 \text{ s}}$.

$v_0 = g t_{up} = (9.81 \text{ m/s}^2)(1.4 \text{ s}) = 14$ m/s. This is the initial *speed*. The initial *velocity* is $\underline{14 \text{ m/s}}$

up.

2-85.   Take the coordinate axis to point up. Then the final displacement is −9.2 m, $a = -g$, $t = 2.5$ s.

$d = v_0 t + \dfrac{at^2}{2} \Rightarrow v_0 = \dfrac{d - \dfrac{at^2}{2}}{t} = \dfrac{d}{t} - \dfrac{at}{2} = \dfrac{-9.2 \text{ m}}{2.5 \text{ s}} - \dfrac{(-9.81 \text{ m/s}^2)(2.5 \text{ s})}{2} = 8.57$ m/s, or $\underline{8.6}$

m/s (to two significant figures). The height the penny reaches above its launch point is

$h = -\dfrac{v_0^2}{2a} = -\dfrac{v_0^2}{2(-g)} = \dfrac{v_0^2}{2g} = \dfrac{(8.57 \text{ m/s})^2}{2(9.81 \text{ (m/s}^2))} = \underline{3.7 \text{ m}}.$

2-87. Assume that the collision of the ball with the floor simply reverses the direction of the ball's velocity. Then the time for the ball to reach the floor is the same as the time for the ball to return to its starting point, and its speed upon returning will be the same as the speed with which it started. Thus $t_{down} = \dfrac{0.90 \text{ s}}{2} = 0.45$ s. Since the ball begins by moving down, choose the direction of the axis for the motion to point down. Then the acceleration and initial velocity are represented by positive numbers. $h = v_0 t + \dfrac{1}{2} gt^2$

$\Rightarrow v_0 = \dfrac{2h - gt^2}{2t} = \dfrac{2(1.5 \text{ m}) - (9.81 \text{ m/s}^2)(0.45 \text{ s})^2}{2(0.45 \text{ s})} = 1.13$ m/s, or $\underline{1.1 \text{ m/s}}$ to two significant figures. The velocity just before hitting the floor is $v = v_0 + gt = $

1.13 m/s + (9.81 m/s^2)(0.45 s) = $\underline{5.5 \text{ m/s}}$.

2-93. The muzzle speed is given by $v_0 = \sqrt{2gh} = \sqrt{2(9.81 \text{ m/s}^2)(180 \times 10^3 \text{ m})} = \underline{1.88 \times 10^3 \text{ m/s}}$. To find out how long the projectile remains above 100 km, we can use the fact that the time for the projectile to climb from 100 km to 180 km is the same as the time for it to fall from 180 km back to 100 km. So we can just calculate the time to fall a distance of 80 km from rest and double that

value. The time to fall can be calculated from $y = \dfrac{gt^2}{2} \Rightarrow t = \sqrt{\dfrac{2y}{g}} = \sqrt{\dfrac{2(80 \times 10^3 \text{ m})}{9.81 \text{ m/s}^2}} = 128$s,

so the total time above 100 km is $\underline{2t = 256 \text{ s}}$.

2-97. $t_{up} = 1.0$ s, $h = 10$ m, $a = -g = 9.81$ m/s^2.

$h = v_0 t + \dfrac{1}{2} at^2 \Rightarrow v_0 = \dfrac{h - \dfrac{at^2}{2}}{t} = \dfrac{10 \text{ m} - \dfrac{(-9.81 \text{ m/s}^2)(1.0 \text{ s})^2}{2}}{1.0 \text{ s}} = \underline{14.9 \text{ m/s}}$. The impact speed is

$v = v_0 + at = 14.9 \text{ m/s} - (9.81 \text{ m/s}^2)(1.0 \text{ s}) = \underline{5.1 \text{ m/s}}$.

2-101. At t = 1 s, $a = \dfrac{dv}{dt} = a_0 (1 - \dfrac{t^2}{4.0 \text{ s}^2})$

$v = \int_0^v a\,dt = \int_0^{1 \text{ s}} a_0 (1 - \dfrac{t^2}{4.0 \text{ s}^2})\, dt = a_0 (t - \dfrac{t^3}{12.0 \text{ s}^2}) \Big|_0^{1 \text{ s}} = 20 \text{ m/s}^2 (1 - \dfrac{1}{12})$ s = $\underline{18.3 \text{ m/s}}$.

After 2 s, the acceleration becomes zero, so the velocity becomes constant at whatever value it had at $t = 2$ s. So to find v after a long time ($t \gg 2$ s), find its value at 2 s:

$v = \int_0^v a\,dt = \int_0^{2 \text{ s}} a_0 (1 - \dfrac{t^2}{4.0 \text{ s}^2})\, dt = a_0 (t - \dfrac{t^3}{12.0 \text{ s}^2}) \Big|_0^{2 \text{ s}} = 20 \text{ m/s}^2 (2 - \dfrac{8}{12})$ s = $\underline{26.7 \text{ m/s}}$

The distance traveled is

$x - x_0 = \int_0^{2 \text{ s}} v\,dt = \int_0^{2 \text{ s}} a_0 (t - \dfrac{t^3}{12.0 \text{ s}^2})dt = a_0 (\dfrac{t^2}{2} - \dfrac{t^4}{48 \text{ s}^2}) \Big|_0^{2 \text{ s}} = (20 \text{ m/s}^2) \left[\dfrac{(2 \text{ s})^2}{2} - \dfrac{(2 \text{ s})^4}{48 \text{ s}^2} \right]$

= $\underline{33.3 \text{ m}}$.

2-103. $a = \dfrac{dv}{dt} = g - Av \Rightarrow \dfrac{dv}{g - Av} = dt \Rightarrow \displaystyle\int_{v_0}^{v} \dfrac{dv}{g - Av} = t$

$\Rightarrow -\dfrac{1}{A} \ln \dfrac{g - Av}{g - Av_0} = t \Rightarrow \dfrac{g - Av}{g - Av_0} = e^{-At} \Rightarrow g - Av = (g - Av_0)e^{-At}$

$\Rightarrow v = \dfrac{1}{A}[g - (g - Av_0)e^{-At}] = \underline{\dfrac{g}{A}(1 - e^{-At}) + v_0 e^{-At}}$

After a long time, $t >> 1/A$, $e^{-At} \to 0$, $v \to \dfrac{g}{A}$.

2-105. Cram's speed is $v_C = \dfrac{1 \text{ mile}}{3 \text{ min } 46.32 \text{ s}} = \dfrac{1 \text{ mile}}{226.32 \text{ s}}$. The time difference for the two runners

getting to the finish line is $\Delta t = 3 \text{ min } 46.32 \text{ sec } - 3 \text{ min } 44.39 \text{ sec } = 1.93$ s. So Cram is behind

by a distance of $v_C \Delta t = \dfrac{1 \text{ mile}}{226.32 \text{ s}} \times 1.93 \text{ s} = 8.53 \times 10^{-3} \text{ mile} \times \dfrac{1609 \text{ m}}{\text{mile}} = \underline{13.7 \text{ m}}.$

2-109. Define: t = time from when the sailfish spots the mackerel to when it catches the mackerel. Then: distance for the sailfish = 109 km/hr $\times t$, distance for the mackerel = 33 km/hr $\times t$. The

separation between the fish is 20 m, so the time for the sailfish to catch the mackerel is given by 109 km/hr $\times t - 33$ km/hr $\times t = 20$ m

$\Rightarrow t = \dfrac{20 \text{ m}}{76 \text{ km/hr}} = \dfrac{20 \text{ m} \cdot 3600 \text{ s/hr}}{76 \text{ km/hr} \cdot 1000 \text{ m/km}} = \underline{0.95 \text{ s}}.$ During this time, the sailfish travels a

distance $d = 109$ km/hr $\times 0.95$ s $= \dfrac{109 \text{ km/hr} \times 1000 \text{ m/km} \times 0.95 \text{ s}}{3600 \text{ s/hr}} = \underline{28.8 \text{ m}}.$

2-111. Distance that my car travels = 80 km/hr $\cdot t$ Distance that the other car travels = 50 km/hr $\cdot t$. To go from 10 m behind the slower car to 10 m ahead of it requires traveling a total relative distance of 10 m + 10 m + 4 m, because of the length of the car. Thus 80 km/hr $\cdot t - 50$ km/hr $\cdot t = (10 + 10 +$

4) m $\Rightarrow t = \dfrac{24 \text{ m}}{30 \text{ km/hr}} = \dfrac{24 \text{ m} \cdot 3600 \text{ s/hr}}{30 \text{ km/hr} \cdot 1000 \text{ m/km}} = \underline{2.9 \text{ s}}.$

2-113. (a) $x = 2.0 + 6.0t - 3.0t^2$. At t = 0.50 s, $x = 2.0 + 6.0 \times 0.50 - 3.0 \cdot (0.50)^2 = \underline{4.3 \text{ m}}.$

(b) $v = \dfrac{dx}{dt} = 6.0 - 6.0t$. At t = 5.0 s, $v = 6.0 - 6.0 \cdot 0.50 = \underline{3.0 \text{ m/s}}$

(c) $a = \dfrac{dv}{dt} = \dfrac{d}{dt}(6.0 - 6.0t) = \underline{-6.0 \text{ m/s}^2 \text{ at all times}}.$

2-115. The initial speed of the car is $v_0 = 90$ km/hr $= 25.0$ m/s. The distance traveled during the reaction time $t_1 = 0.75$ s is $d_1 = v_0 t_1 = (25 \text{ m/s})(0.75 \text{ s}) = 18.8$ m. The remaining distance to the cow is d_2 = 30 m – 18.8 m = 11.2 m. The car's acceleration as it travels this distance is $a = -8.0$ m/s^2. Its final speed when it hits the cow is given by

$v^2 - v_0^2 = 2ad_2 \Rightarrow v = \sqrt{v_0^2 + 2ad_2} = \sqrt{(25 \text{ m/s})^2 + 2(-8.0 \text{ m/s})(11.2 \text{ m})} = 21.1$ m/s, or

$\underline{76 \text{ km/hr}}.$

2-119.   (a) $\Delta t = 1$ s. The distance the first ball falls in that interval is $y_1 = \dfrac{1}{2} g t^2$

$= \dfrac{1}{2}(9.81 \text{ m/s}^2)(1 \text{ s})^2 = 4.90$ m, so the first ball is 13 m $- 4.9$ m $= \underline{8.1 \text{ m above the ground}}$ when

the second ball is released. The time for the first ball to fall the total distance of 13 m is

$t_1 = \sqrt{\dfrac{2h}{g}} = \sqrt{\dfrac{2 \times 13 \text{ m}}{9.81 \text{ m/s}^2}} = 1.63$ s. When the first ball hits the ground, the second ball has been

falling for $t_2 = 0.63$ s. The distance second ball has fallen during this time is

$y_2 = \dfrac{1}{2} g t_2^2 = \dfrac{1}{2}(9.81 \text{ m/s}^2)(0.63 \text{ s})^2 = 1.95$ m, so the second ball is 13 m $- 1.95$ m $= \underline{11.1 \text{ m}}$

above the ground when the first ball lands.

(b) $v (1\text{st}) = g t_1 = 9.81$ m/s$^2 \cdot 1.63$ s $= 16.0$ m/s

$v(2\text{nd}) = g t_2 = 9.81$ m/s$^2 \cdot 0.63$ s $= 6.18$ m/s

$\Rightarrow$ instantaneous velocity of the first ball relative to the second just before the first hits the ground
is: 16.0 m/s $- 6.18$ m/s $= \underline{9.8 \text{ m/s down}}$.

(c) Both balls have same acceleration, (9.81 m/s^2 down,) so the relative acceleration is $\underline{\text{zero}}$.

CHAPTER 3   VECTORS

3-3.   A reduced copy of the diagram is shown. In the actual
diagram, 1 cm = 1 km. From the diagram, the line
representing the resultant **R** is 11.2 cm long, so the length
of **R** is 11.2 km. The angle θ is measured to be 27.5°.
Graphically we find **R** = <u>11.2 km @ 27.5° S of E</u>.
To do the problem trigonometrically, take x to point east
and y to point north. Then

$$\mathbf{r}_1 = 18.0\sin 60°\mathbf{i} + 18.0\cos 60°\mathbf{j} = 15.6\mathbf{i} + 9.0\mathbf{j}\ \text{km}$$

$$\mathbf{r}_2 = 9.5\cos 60°\mathbf{i} - 9.5\sin 60°\mathbf{j} = 4.75\mathbf{i} - 8.23\mathbf{j}\ \text{km}$$

$$\mathbf{r}_3 = -12.0\sin 60°\mathbf{i} - 12.0\cos 60°\mathbf{j} = -10.4\mathbf{i} - 6.0\mathbf{j}\ \text{km}$$

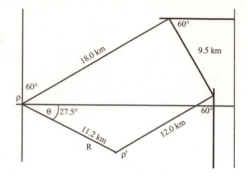

Adding gives **R** = $9.95\mathbf{i} - 5.23\mathbf{j}$ km. The magnitude is $R = \sqrt{9.95^2 + 5.23^2} = 11.2$ km, and the

direction is $\theta = \tan^{-1}\dfrac{-5.23}{9.95} = -27.7°$. In terms of compass directions, this is **R** = <u>11.2 km @</u>

<u>27.7° S of E</u>.

3-5.

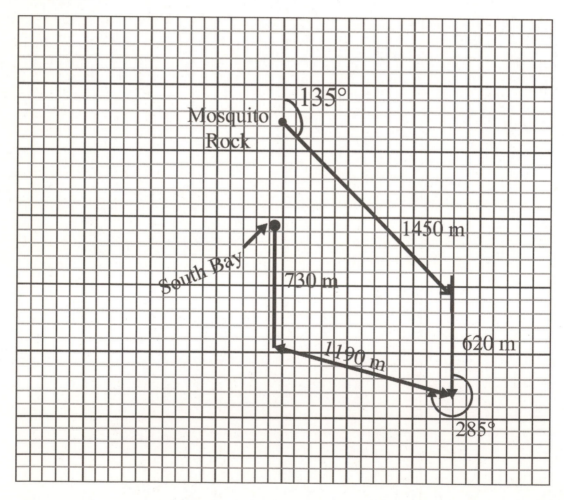

14

The graph is drawn with a scale of 1 cm = 20 m. Measuring the distance from South Bay to Mosquito Rock, we get: $\mathbf{r} \approx -0.6$ cm $\mathbf{i} - 3.0$ cm $\mathbf{j}$ on the graph $= -120$ m $\mathbf{i} - 600$ m $\mathbf{j}$.

The distance between them is $\sqrt{(-120)^2 + (-600)^2} = \underline{612}$ m at an angle $\theta = \tan^{-1} \dfrac{120}{600} = $

$\underline{11.3°}$ west of south.

3-7.   A reduced copy of the graphical solution is shown in the diagram. In the actual graph, the scale is 1.0 cm = 100 km, so the line for the first displacement is 4.8 cm long and the line for the second displacement is 3.7 cm long. The line for the total displacement is measured to be 4.4 cm long, corresponding to a magnitude of 440 km. The direction is measured to be 7.5° W of N. The total displacement is $\mathbf{D} = \underline{440}$ km @ 7.5° W of N.

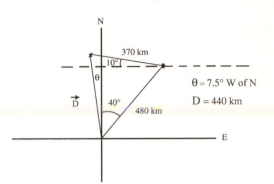

Check analytically. Let $\mathbf{r}_1 = 480 \sin (40°)$ km $\mathbf{i} + 480$ km $\cos (40°)$ $\mathbf{j} = 308.5$ km $\mathbf{i} + 367.7$ km $\mathbf{j}$ and $\mathbf{r}_2 = -370 \cos (10°)$ km $\mathbf{i} + 370$ km $\cos (10°)$ $\mathbf{j} = -364.4$ km $\mathbf{i} + 64.2$ km $\mathbf{j}$, where E corresponds to x and N corresponds to y. Then $\mathbf{D} = \mathbf{r}_1 + \mathbf{r}_2 = -55.9$ km $\mathbf{i} + 431.9$ km $\mathbf{j}$.

$|\mathbf{D}| = \sqrt{(-55.9)^2 + 431.9^2}$ km $= \underline{436 \text{ km}}$. To get the direction relative to the x direction, note that the resultant vector is in the third quadrant (negative x component, positive y component). The standard method of finding the direction will give an angle measured from E, the x axis. Using a calculator is used to find the inverse tangent will give a negative angle, so the correct angle (in the second quadrant) is found by adding 180° to the calculator result:

$\theta = 180° - \tan^{-1} \dfrac{431.9}{55.9} = 97.4°$, which measured N of E. This is 7.4° W of N. The analytical

result is $\mathbf{D} = \underline{436 \text{ km}}$ @ 7.4° W of N, which agrees almost exactly with the graphical result.

3-9.   The resultant displacement vector, $\mathbf{R} = (5$ m). The given displacement vector, $\mathbf{A} = (2.2$ m) $\sin 35°\,\mathbf{i} + (2.2$ m) $\cos 35°\,\mathbf{j} = (1.26$ m) $\mathbf{i} + (1.8$ m)$\mathbf{j}$. Let the other displacement vector be $\mathbf{B} = B_x\mathbf{i} + B_y\mathbf{j}$. Therefore, $\mathbf{R} = \mathbf{A} + \mathbf{B} = (1.26$ m $+ B_x)\,\mathbf{i} + (1.8$ m $+ B_y)\,\mathbf{j} = 5$ m$\mathbf{j}$. Comparing $\mathbf{i}$ and $\mathbf{j}$ components gives

1.26 m $+ B_x = 0$ or $B_x = -1.26$ m
1.8 m $+ B_y = 5$ or $B_y = 3.2$ m
Thus, $\mathbf{B} = \underline{(-1.26 \text{ m})\mathbf{i} + (3.2 \text{ m})\mathbf{j}}$.

3-11.   The easiest method is to write the vectors in component form, with $\hat{\mathbf{j}}$ as N and $\hat{\mathbf{i}}$ as E.

Then $\mathbf{R} = 1.2\,\hat{\mathbf{j}} + (\sin 38°\,\hat{\mathbf{i}} \ \cos 38°\,\hat{\mathbf{j}})6.1 + (\sin 59°\,\hat{\mathbf{i}} - \cos 59°\,\hat{\mathbf{j}})2.9 +$

$(\sin 89°\,\hat{\mathbf{i}} + \cos 89°\,\hat{\mathbf{j}})\,4.0 + (\sin 31°\,\hat{\mathbf{i}} - \cos 31°\,\hat{\mathbf{j}})6.5$

$= 1.2\,\hat{\mathbf{j}} + 3.76\,\hat{\mathbf{i}} - 4.81\,\hat{\mathbf{j}} + 2.49\,\hat{\mathbf{i}} - 1.49\,\hat{\mathbf{j}} + 4.00\,\hat{\mathbf{i}} + 0.07\,\hat{\mathbf{j}} + 3.35\,\hat{\mathbf{i}} + 5.57\,\hat{\mathbf{j}}$

$= 13.6\,\hat{\mathbf{i}} + 0.54\,\hat{\mathbf{j}}$

This translates into a vector of length $\underline{13.6}$ nmi at $\underline{88°}$ E of N.

3-19. Place the x axis along the sloping line as shown. Then: $\underline{x}$
$\underline{\text{component} = -(4.0\text{ m})\sin(25°) = -1.7\text{ m}}$.

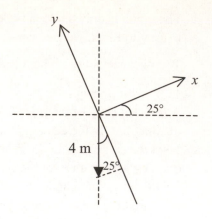

3-21. (a)

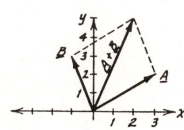

(b) $\mathbf{A} = 3\mathbf{i} + 2\mathbf{j}$ cm
$\mathbf{B} = -1\mathbf{i} + 3\mathbf{j}$ cm
$\mathbf{A} + \mathbf{B} = 3\mathbf{i} + 2\mathbf{j} - 1\mathbf{i} + 3\mathbf{j}$ cm
$\underline{= (2\mathbf{i} + 5\mathbf{j}\text{ cm})}$

3-23. $x : 6.0\cos 45° = \underline{4.2\text{ units}}$

$y : 6.0\cos 85° = \underline{0.5\text{ units}}$

Since the magnitude of the vector is 6.0 units, we know that
$A = \sqrt{A_x^2 + A_y^2 + A_z^2} = 6.0^2$. Thus $A_z = \pm\sqrt{6.0^2 - 4.2^2 - 0.5^2} = \pm 4.2$ units

$\underline{A_z \text{ can be either} \pm 4.2 \text{ units, so it is not uniquely determined.}}$

3-29. $\hat{\mathbf{r}} = \dfrac{\mathbf{A}}{|\mathbf{A}|} = \dfrac{2\mathbf{i} + 4\mathbf{j} + 4\mathbf{k}}{\sqrt{2^2 + 4^2 + 4^2}} = \dfrac{1}{3}\mathbf{i} + \dfrac{2}{3}\mathbf{j} + \dfrac{2}{3}\mathbf{k}$

3-31. $c_1\mathbf{A} + c_2\mathbf{B} = c_1(2.0\,\mathbf{i} + 3.0\,\mathbf{j}) + c_2(1.0\,\mathbf{i} + 5.0\,\mathbf{j})$
$= (2.0\,c_1 + 1.0c_2)\mathbf{i} + (3.0\,c_1 + 5.0c_2)\mathbf{j}$
$c_1\mathbf{A} + c_2\mathbf{B} = \mathbf{C} = -1.0\,\mathbf{i} + 3.0\,\mathbf{j}$
$\Rightarrow 2.0c_1 + 1.0c_2 = -1.0$ (1)
$3.0c_1 + 5.0c_2 = 3.0$ (2)

Solve for c_1, c_2: (1) $\times 3 - $ (2) $\times 2 : \Rightarrow -7c_2 = -9 \Rightarrow \underline{c_2 = \dfrac{9}{7}}, \underline{c_1 = -\dfrac{8}{7}}$

3-35. $3\mathbf{i} - 6\mathbf{j} + 2\mathbf{k}$ has magnitude $\sqrt{3^2 + 6^2 + 2^2} = \sqrt{49} = 7$ units

The unit vector in this direction is $\dfrac{1}{7}(3\mathbf{i} - 6\mathbf{j} + 2\mathbf{k})$

Therefore the vector with magnitude $\underline{2}$ is $2\times$ (unit vector)
$= \underline{\dfrac{6}{7}\mathbf{i} - \dfrac{12}{7}\mathbf{j} + \dfrac{4}{7}\mathbf{k}}$

16

3-39. $A = \sqrt{2^2 + 1^2 + 2^2} = 3$

$B = \sqrt{3^2 + 6^2 + 2^2} = 7$

Dot product $\mathbf{A} \cdot \mathbf{B} = (-2)(3) + (1)(-6) + (2)(2) = \underline{-8}$

$\mathbf{A} \cdot \mathbf{B} = AB \cos \phi$

$\phi = \cos^{-1} \dfrac{\mathbf{A} \cdot \mathbf{B}}{A \, B} = \cos^{-1} \dfrac{8}{7 \times 3} = \underline{112°}$

3-41. $\cos \theta = \dfrac{\mathbf{A} \cdot \mathbf{i}}{|\mathbf{A}|} = \dfrac{3 \cdot 1 + 4 \cdot 0 + 2 \cdot 0}{\sqrt{3^2 + 4^2 + 2^2}} = 0.557 \Rightarrow \theta = \cos^{-1} 0.557 = \underline{56.1°}$

3-43. $\mathbf{A} \cdot \mathbf{B} = |\mathbf{A}| \, |\mathbf{B}| \cos \theta$

$\mathbf{A} \times \mathbf{B} = |\mathbf{A}| \, |\mathbf{B}| \sin \theta$

If $|\mathbf{A} \times \mathbf{B}| = \mathbf{A} \cdot \mathbf{B}$, then $|\mathbf{A}| \, |\mathbf{B}| \cos \theta = |\mathbf{A}| \, |\mathbf{B}| \sin \theta \Rightarrow \cos \theta = \sin \theta$

which gives: $\theta = \underline{45°}$

3-45. Take north to be $\mathbf{j}$ and east to be $\mathbf{i}$. Then

$\mathbf{A} \times \mathbf{B} = (2180\mathbf{i}) \times (-1790\mathbf{j}) \, \text{km}^2 = -3.90 \times 10^6 (\mathbf{i} \times \mathbf{j}) \, \text{km}^2 = \underline{-3.90 \times 10^6 \mathbf{k} \, \text{km}^2}$

3-49. $\mathbf{A} \times \mathbf{B} = (A_y B_z - A_z B_y)\mathbf{i} + (A_z B_x - A_x B_y)\mathbf{j} + (A_x B_y - A_y B_x)\mathbf{k}$

$\mathbf{A} \times \mathbf{B} = (5.0\mathbf{i} - 2.0\mathbf{j} + 3.0\mathbf{k}) \times (B_x\mathbf{i} + 3.0\mathbf{j} + B_z\mathbf{k}) = (-2B_z - 9)\mathbf{i} + (3B_x - 5B_z)\mathbf{j} + (15 + 2B_x)\mathbf{k}$

$\mathbf{A} \times \mathbf{B} = \mathbf{C} = 2.0\mathbf{j} + C_z\mathbf{k}.$

$(\mathbf{A} \times \mathbf{B})_x = A_y B_z - A_z B_y = C_x \Rightarrow -2B_z - 9 = 0 \Rightarrow \underline{B_z = -4.5}$

$(\mathbf{A} \times \mathbf{B})_y = A_z B_x - A_x B_z = C_y \Rightarrow 3B_x - 5B_z = 2 \Rightarrow B_x = \dfrac{1}{3}(2 + 5B_z)$

$= \dfrac{1}{3}[2 + 5 \times (-4.5)] = \underline{-6.83}.$

$(\mathbf{A} \times \mathbf{B})_z = A_x B_y - A_y B_x = C_z \Rightarrow 15 + 2B_x = C_z \Rightarrow C_z = 15 + 2(-6.83) = \underline{1.34}$

3-55. Since $|\mathbf{A}| \neq 0$, $|\mathbf{B}| \neq 0$, yet $\mathbf{A} \cdot \mathbf{B} = 0$, we can conclude the angle between $\mathbf{A}$ and $\mathbf{B}$ is 90°.

So $|\mathbf{A} \times \mathbf{B}| = |\mathbf{A}| \, |\mathbf{B}| \sin 90° = 4 \times 6 = \underline{24}$

3-59. $\mathbf{A} = 2\mathbf{i} - 3\mathbf{j} + 2\mathbf{k}.$ $\mathbf{B} = -3\mathbf{i} + 4\mathbf{k}.$ $\mathbf{A} \times \mathbf{B} = (2\mathbf{i} - 3\mathbf{j} + 2\mathbf{k}) \times (-3\mathbf{i} + 0\mathbf{j} + 4\mathbf{k})$

$= (A_y B_z - A_z B_y)\mathbf{i} + (A_z B_x - A_x B_z)\mathbf{j} + (A_x B_y - A_y B_x)\mathbf{k}$

$= [(-3)(4) - (2)(0)]\mathbf{i} + [(2)(-3) - (2)(4)]\mathbf{j} + [(2)(0) - (-3)(-3)]\mathbf{k} = \underline{-12\mathbf{i} - 14\mathbf{j} - 9\mathbf{k}}.$

3-67. By the law of cosines:

$C^2 = A^2 + B^2 - 2AB \cos 115°$

$= 350^2 + 120^2 + 2(350)(120)(0.4226)$

$= 172,400 \, \text{m}^2$

$\underline{C = 415 \, \text{m}}$

By the law of sines, the angle between A and C is given by

$\dfrac{\sin \theta}{120} = \dfrac{\sin 115°}{415}$

$\theta = \sin^{-1}\left(\dfrac{120}{415} \sin 115°\right) = \underline{15.2°}$

Therefore, the angle between C and north is 45° −

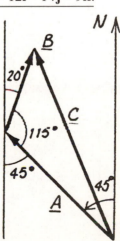

15.2° = 29.8° and the <u>resultant has a magnitude of</u>
<u>415 m, at 29.8° W of N.</u>

3-71. $\mathbf{A} = 6.2 \cos 30°\mathbf{i} - 6.2 \sin 30°\mathbf{j}$

$\mathbf{B} = -9.6\mathbf{j}$

Therefore, $\mathbf{A} + \mathbf{B} = 6.2 \cos 30°\mathbf{i} + (-6.2 \sin 30° - 9.6)\mathbf{j} = \underline{5.4\mathbf{i} - 12.7\mathbf{j}}$

and $\mathbf{A} - \mathbf{B} = 6.2 \cos 30°\mathbf{i} - (6.2 \sin 30° - 9.6)\mathbf{j} = \underline{5.4\mathbf{i} + 6.5\mathbf{j}}$

3-77.   If ϕ is the angle between $\mathbf{A}$ and $\mathbf{B}$, then the component of $\mathbf{A}$ along $\mathbf{B}$ is $A \cos \phi$. But $\cos \phi =$

$(\mathbf{A} \cdot \mathbf{B})/(AB)$. Therefore $A \cos \phi = A\left(\dfrac{\mathbf{A} \cdot \mathbf{B}}{AB}\right) = \dfrac{\mathbf{A} \cdot \mathbf{B}}{B}$. $B = \sqrt{1^2 + 3^2 + 2^2} = \sqrt{14} = 3.74$.

$\mathbf{A} \cdot \mathbf{B} = (3)(1) + (4)(3) + (0)(-2) = 15$. Then $A \cos \phi = \dfrac{15}{3.74} = \underline{4.0}$.

Similarly, the component of $\mathbf{B}$ along $\mathbf{A}$ is $\dfrac{\mathbf{A} \cdot \mathbf{B}}{A}$ with $A = \sqrt{3^2 + 4^2} = 5$. Thus, $B \cos \phi$

$= \dfrac{15}{5} = \underline{3.0}$.

4-7.   At $t = 2.0$ s, the missile has been falling with acceleration $\mathbf{a} = 0\mathbf{i} - (9.81 \text{ m/s}^2)\mathbf{j}$ and horizontal velocity equal to the velocity of the airplane. This means the missile is still directly below the airplane. Its displacement relative to the plane is $\mathbf{r} = 0\mathbf{i} - \dfrac{gt^2}{2}\mathbf{j} = 0\mathbf{i} - \dfrac{(9.81 \text{ m/s}^2)(2.0 \text{ s})^2}{2}\mathbf{j} =$
$0\mathbf{i} - (19.6 \text{ m})\mathbf{j}$, or $\underline{19.6 \text{ m @ } 90^\circ \text{ below the direction of travel of the airplane}}$. At $t = 3.0$ s, which is 1.0 s after igniting the engine, the acceleration is $\mathbf{a} = (6.0 \text{ m/s}^2)\mathbf{i} - (9.81 \text{ m/s}^2)\mathbf{j}$. Its
displacement during this 1.0 s interval is $\mathbf{r}_1 = \dfrac{a_x t^2}{2}\mathbf{i} - \dfrac{gt^2}{2}\mathbf{j}$
$= \dfrac{(6 \text{ m/s}^2)t^2}{2}\mathbf{i} - \dfrac{gt^2}{2}\mathbf{j} = \dfrac{(6 \text{ m/s}^2)(1.0 \text{ s})^2}{2}\mathbf{i} - \dfrac{(9.81 \text{ m/s}^2)(1.0 \text{ s})^2}{2}\mathbf{j} = (3.0 \text{ m})\mathbf{i} - (4.9 \text{ m})\mathbf{j}$. Now
the missile's total displacement relative to the plane is
$\mathbf{r}_2 = (0 \text{ m})\mathbf{i} - (19.6 \text{ m})\mathbf{j} + (3.0 \text{ m})\mathbf{i} - (4.9 \text{ m})\mathbf{j} = (3.0 \text{ m})\mathbf{i} - (24.5 \text{ m})\mathbf{j}$. The magnitude is
$r_2 = \sqrt{(3.0 \text{ m})^2 + (24.5 \text{ m})^2} = 24.7$ m. The direction is given by $\theta = \tan^{-1}\dfrac{-24.5}{3} = -83.0^\circ$, or
$\underline{24.7 \text{ m @ } 83^\circ \text{ below the direction of travel of the plane}}$.

4-9. $\mathbf{v}_{average} = \dfrac{\Delta \mathbf{r}}{\Delta t}$, where $\Delta \mathbf{r}$ is the total displacement and $\Delta t = 1.5$ h is the total elapsed time. To find
the total displacement, find the displacement during each part of the trip:
$\Delta \mathbf{r}_1 = (300 \text{ km/h @ } 30^\circ \text{ N of E}) \times 0.50 \text{ h} = 150 \text{ km @ } 30^\circ \text{ N of E}$. Taking the y direction to
point N and the x direction to point E, this is $\Delta \mathbf{r}_1 = (150 \text{ km})(\cos 30^\circ)\mathbf{i} + (150 \text{ km})(\sin 30^\circ)\mathbf{j}$
$= (130 \text{ km})\mathbf{i} + (75 \text{ km})\mathbf{j}$. For the second part of the trip,
$\Delta \mathbf{r}_2 = (300 \text{ km/h @ } 30^\circ \text{ W of S}) \times 1.0 \text{ h} = 300 \text{ km @ } 30^\circ \text{ W of S}$. In terms of x and y, this
is $\Delta \mathbf{r}_2 = -(300 \text{ km})(\sin 30^\circ)\mathbf{i} - (300 \text{ km})(\cos 30^\circ)\mathbf{j} = -(150 \text{ km})\mathbf{i} - (260 \text{ km})\mathbf{j}$. The total
displacement is $\Delta \mathbf{r} = \Delta \mathbf{r}_1 + \Delta \mathbf{r}_2 = -20 \text{ km } \mathbf{i} - 185 \text{ km } \mathbf{j}$. The average velocity is $\mathbf{v}_{average} = \dfrac{\Delta \mathbf{r}}{\Delta t}$
$= \dfrac{-20 \text{ km } \mathbf{i} - 185 \text{ km } \mathbf{j}}{1.5 \text{ h}} = -13.3 \text{ km/h } \mathbf{i} - 123 \text{ km/h } \mathbf{j}$. To give this as a
speed and heading, use $v_{average} = \sqrt{13.3^2 + 123^2} \text{ km/h} = 124$ km/h. Since
both velocity components are negative, the vector is located in the third
quadrant (between W and S), so we can give the direction
as $\theta = \tan^{-1}\dfrac{123}{13.3} = 83.8^\circ$ S of W. Thus $\mathbf{v}_{average} = \underline{124 \text{ km/h @ } 83.8^\circ \text{ S of}}$
$\underline{\text{W}}$.

The average acceleration is $\mathbf{a}_{average} = \dfrac{\Delta \mathbf{v}}{\Delta t} = \dfrac{\mathbf{v}_2 - \mathbf{v}_1}{\Delta t}$, where $\mathbf{v}_1$ is the velocity during the 0.5, $\mathbf{v}_2$ is
the velocity during the next 1.0 h, and Δt is the time interval during which the velocity changes.
Since no value is given for Δt, we can calculate $\Delta \mathbf{v}$ and give the answer symbolically. Using the
definitions above, $\mathbf{v}_1 = (300 \text{ km/h})\cos 30^\circ \mathbf{i} + (300 \text{ km/h})\sin 30^\circ \mathbf{j} = (260 \text{ km/h})\mathbf{i} + (150 \text{ km/h})\mathbf{j}$.
30° W of S is the same as 240° N of E, so $\mathbf{v}_2 = (300 \text{ km/h})\cos 240^\circ \mathbf{i} + (300 \text{ km/h})\sin 240^\circ \mathbf{j}$
$= -(150 \text{ km/h})\mathbf{i} - (240 \text{ km/h})\mathbf{j}$. Then $\Delta \mathbf{v} = -(410 \text{ km/h})\mathbf{i} - (410 \text{ km/h})\mathbf{j}$,

and $\mathbf{a}_{average} = \dfrac{-(410 \text{ km/h})\mathbf{i} - (410 \text{ km/h})\mathbf{j}}{\Delta t}$. As a magnitude and heading, this

is $\mathbf{a}_{average} = \dfrac{410\sqrt{2} \text{ km/h}}{\Delta t}$ @ 45° S of W, or $\mathbf{a}_{average} = \dfrac{580 \text{ km/h}}{\Delta t}$ @ 45° S of W. Even though we

don't have a numerical value, it is important to note that the average acceleration is *not* zero, even though the speed of the airplane is constant. The acceleration is caused by the change in direction of the velocity.

4-11. $\mathbf{r} = 90t\mathbf{i} + (500 - 15t)\mathbf{j}$ m. $\mathbf{v} = \dfrac{d\mathbf{r}}{dt} = 90\mathbf{i} - 15\mathbf{j}$ m/s. The equation

for the speed is

$v = \sqrt{v_x^2 + v_y^2} = \sqrt{(90 \text{ m/s})^2 + (15 \text{ m/s})^2} = 91$ m/s. The direction

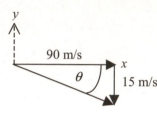

of the velocity is $\theta = \tan^{-1}\dfrac{v_y}{v_x} = \tan^{-1}\left(\dfrac{-15}{90}\right) = -9.5°$.

This is 9.5° below the *x* axis.

4-13. $\mathbf{a} = \dfrac{d\mathbf{v}}{dt} = 3\mathbf{i} + 2\mathbf{j} \Rightarrow d\mathbf{v} = (3\mathbf{i} + 2\mathbf{j})dt$. Integrate to find $\mathbf{v}$: $\displaystyle\int_{v_0}^{v} d\mathbf{v} = \int_0^t (3\mathbf{i} + 2\mathbf{j})dt$, which gives

$\mathbf{v} - \mathbf{v}_0 = 3t\mathbf{i} + 2t\mathbf{j}$. The problem states that both components of velocity are initially zero, so we

get $\mathbf{v} = 3t\mathbf{i} + 2t\mathbf{j}$ m/s. Since $\mathbf{v} = \dfrac{d\mathbf{r}}{dt}$, we can follow the same procedure that was used to get $\mathbf{v}$

from $\mathbf{a}$: $d\mathbf{r} = (3t\mathbf{i} + 2t\mathbf{j})dt \Rightarrow \displaystyle\int_{r_0}^{r} d\mathbf{r} = \int_0^t (3t\mathbf{i} + 2t\mathbf{j})dt \Rightarrow \mathbf{r} - \mathbf{r}_0 = \dfrac{3t^2}{2}\mathbf{i} + t^2\mathbf{j}$. The problem says

the particle starts moving at the origin, so both components of $\mathbf{r}_0$ are zero. $\mathbf{r} = \dfrac{3t^2}{2}\mathbf{i} + t^2\mathbf{j}$ m.

4-15. Assume the divers launch themselves horizontally from the edge of the cliff, so $v_{oy} = 0$. The time

to fall a distance $h = 64$ m is $t = \sqrt{\dfrac{2h}{g}} = \sqrt{\dfrac{2(36 \text{ m})}{9.81 \text{ m/s}^2}} = 2.71$ s. During this time, the diver must

travel a horizontal distance of at least $x = 6.4$ m, so the minimum horizontal velocity required

is $v_{0x} = \dfrac{x}{t} = \dfrac{6.4 \text{ m}}{2.71 \text{ s}} = 2.4$ m/s.

4-17. The stunt car will fall 2 m while it travels 24 m horizontally.

The time to fall 2 m is $t = \sqrt{\dfrac{2h}{g}} = \sqrt{\dfrac{2(2 \text{ m})}{9.81 \text{ m/s}^2}} = 0.639$ s. The

car must travel 24 m horizontally in 0.639 s, so its horizontal speed must be 38 m/s. (This is about 135 kph, or 84 mph.)

4-31. $\mathbf{a} = \dfrac{d\mathbf{v}}{dt} = 2.0\mathbf{i} - 4.5\mathbf{j} \Rightarrow d\mathbf{v} = (2.0\mathbf{i} - 4.5\mathbf{j})dt$. Integrate to find $\mathbf{v}$ following the method used in

Problem 4-13: $\mathbf{v} - \mathbf{v}_0 = \displaystyle\int_0^t (2.0\mathbf{i} - 4.5\mathbf{j})dt = 2.0t\mathbf{i} - 4.5t\mathbf{j}$. The problem states

that $\mathbf{v}_0 = -10\mathbf{i} + 25\mathbf{j}$, so we get $\mathbf{v} = 2.0t\mathbf{i} - 4.5t\mathbf{j} + (-10\mathbf{i} + 25)\mathbf{j} = (2.0t - 10)\mathbf{i} + (25 - 4.5t)\mathbf{j}$

m/s.

At $t = 3.0$ s, the velocity is $\mathbf{v} = (6.0 - 10)\mathbf{i} + (25 - 13.5)\mathbf{j} = -4.0\mathbf{i} + 11.5\mathbf{j}$ m/s. The speed is

$$v = \sqrt{v_x^2 + v_y^2} = \sqrt{(4.0 \text{ m/s})^2 + (11.5 \text{ m/s})^2} = \underline{12 \text{ m/s}}.$$

Since $\mathbf{v} = \dfrac{d\mathbf{r}}{dt}$, we can follow the same procedure that was used to get $\mathbf{v}$ from $\mathbf{a}$:

$$d\mathbf{r} = [(2.0t - 10)\mathbf{i} + (25 - 4.5t)\mathbf{j}]dt \Rightarrow \int_{\mathbf{r}_0}^{\mathbf{r}} d\mathbf{r} = \int_0^t [(2.0t - 10)\mathbf{i} + (25 - 4.5t)\mathbf{j}]dt \Rightarrow$$

$\mathbf{r} - \mathbf{r}_0 = (t^2 - 10t)\mathbf{i} + (25t - 2.25t^2)\mathbf{j}$ m. The problem says the particle starts moving at the origin, so both components of $\mathbf{r}_0$ are zero. $\mathbf{r} = \underline{(t^2 - 10t)\mathbf{i} + (25t - 2.25t^2)\mathbf{j} \text{ m}}$. At $t = 3.0$ s,

$\mathbf{r} = -21\mathbf{i} + 55\mathbf{j}$ m.

4-35. The time to reach maximum height is $t_{height} = \dfrac{2.25 \text{ s}}{2} = 1.125$ s. $t_{height} = \dfrac{v_{0y}}{g} \Rightarrow$

$v_{0y} = gt_{height} = (9.81 \text{ m/s}^2)(1.125 \text{ s}) = 11.0$ m/s. If we say the launch point was at $y = 0$, then the

equation for the position below the launch point is $y = v_{0y}t - \dfrac{gt^2}{2}$ with $v_{0y} = 11.0$ m/s. This

gives $y = (11.0 \text{ m/s})(4.00 \text{ s}) - \dfrac{(9.81 \text{ m/s}^2)(4.00 \text{ s})^2}{2} = -34.5$ m. The cliff is $\underline{34.5 \text{ m}}$ high. Since

the launch angle was $45°$, $v_{0x} = v_{0y} = 11.0$ m/s. The total horizontal distance traveled since

launch is $x = v_{0x}t_{total} = (11.0 \text{ m/s})(2.25 \text{ s} + 4.00 \text{ s}) = \underline{68.8 \text{ m}}$.

4-41. $x_{max} = \dfrac{2v_0^2 \sin\theta \cos\theta}{g}$, $y_{max} = \dfrac{v_0^2 \sin^2\theta}{2g}$. Setting these equal to each other gives

and then a miracle occurs?!?

$\dfrac{2v_0^2 \sin\theta \cos\theta}{g} = \dfrac{v_0^2 \sin^2\theta}{2g}$, from which we get $\dfrac{\sin\theta}{\cos\theta} = \tan\theta = 2$, or $\theta = \tan^{-1} 2 = \underline{63.4°}$.

4-43. A rough sketch (not to scale) is shown. The coordinates of the projectile are

$$y = h + v_0(\sin\theta)t - \dfrac{gt^2}{2}$$

$$x = v_0(\cos\theta)t$$

The coordinates of the target are $(x_{max}, 0)$. The flight time

then is $t = \dfrac{x_{max}}{v_0 \cos\theta}$. Set $y = 0$ and substitute this time.

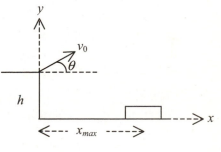

After some rearranging, the equation becomes $g\left(\dfrac{x_{max}^2}{v_0^2 \cos^2\theta}\right) - 2v_0 \sin\theta\left(\dfrac{x_{max}}{v_0 \cos\theta}\right) - 2h = 0$. We

can convert this to a quadratic equation with $\tan\theta$ as the variable by using the following

trigonometric steps: $\tan\theta = \dfrac{\sin\theta}{\cos\theta}$. $\tan^2\theta = \dfrac{\sin^2\theta}{\cos^2\theta} = \dfrac{1 - \cos^2\theta}{\cos^2\theta} = \dfrac{1}{\cos^2\theta} - 1$. Thus

$\dfrac{1}{\cos^2\theta} = 1 + \tan^2\theta$, and we have $\left(\dfrac{gx_{max}^2}{v_0^2}\right)(1 + \tan^2\theta) - 2x_{max}\tan\theta - 2h = 0$. Multiplying by

$\left(\dfrac{v_0^2}{gx_{max}^2}\right)$ and rearranging finally gives this equation:

$\tan^2\theta - 2x_{max}\left(\dfrac{v_0^2}{gx_{max}^2}\right)\tan\theta + 1 - 2h\left(\dfrac{v_0^2}{gx_{max}^2}\right) = 0$. Using $x_{max} = 12 \times 10^3$ m, $v_0 = 600$ m/s, and $g =$

9.81 m/s^2 gives $\left(\dfrac{v_0^2}{gx_{max}^2}\right) = 2.548 \times 10^{-4}$ m^{-1}. Substituting $h = 50$ m and doing the rest of the

arithmetic finally gives $\tan^2\theta - 6.116\tan\theta + 0.9745 = 0$. The solution is

$\tan\theta = \dfrac{6.116 \pm \sqrt{6.116^2 - 4(0.9745)}}{2} = 5.953,\ 0.1635$. The two angles are the inverse tangents

of these: $\theta = \underline{80.5°,\ 9.29°}$. It turns out that most projectile aiming problems like this have two
solutions. The difference between the two angles is that the larger elevation will give a longer
flight time and the projectile will strike the target at a larger angle than for the smaller elevation.

4-49.   Consider the motion of two of the projectiles. Call their positions (x_1, y_1) and (x_2, y_2) and assume
that they were launched with elevation angles θ_1 and θ_2, respectively. The positions are given by

$x_1 = v_0\cos\theta_1 t$, $y_1 = v_0\sin\theta_1 t - \dfrac{gt^2}{2}$ and $x_2 = v_0\cos\theta_2 t$, $y_2 = v_0\sin\theta_2 t - \dfrac{gt^2}{2}$. For the projectiles

to collide, they must have the same positions at the same time: $x_1 = x_2$ and $y_1 = y_2$. This gives the

following pair of equations: $v_0\cos\theta_1 t = v_0\cos\theta_2 t$, $v_0\sin\theta_1 t - \dfrac{gt^2}{2} = v_0\sin\theta_2 t - \dfrac{gt^2}{2}$. Canceling

terms gives $\cos\theta_1 = \cos\theta_2$, $\sin\theta_1 = \sin\theta_2$. Dividing the second equation by the first gives
$\tan\theta_1 = \tan\theta_2$ as the requirement for the projectiles to collide. Since we assumed that the
projectiles were launched at different angles, we arrive at a contradiction, and we conclude that
the individual projectiles will never collide during their flight.

4-51.   Take the x and z directions to point as shown. (The y direction is up,
out of the page.) The projectile has two horizontal components of
velocity: $\mathbf{v}_{horizontal} = v_0\cos\theta\mathbf{i} + v_s\mathbf{k}$, where v_0 is the muzzle speed of
the projectile, θ is the elevation angle of the projectile, and v_s is the
speed of the ship. Neglecting air resistance, the flight time for the

projectile is $t_{flight} = \dfrac{2v_0\sin\theta}{g} = \dfrac{2(720\text{ m/s})(\sin 30°)}{9.81\text{ m/s}^2} = 73.5$ s. The

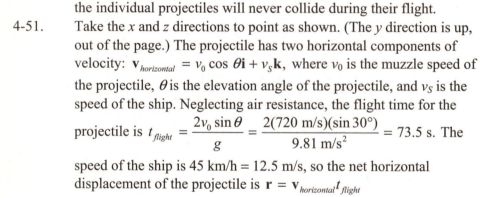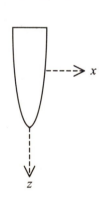

speed of the ship is 45 km/h = 12.5 m/s, so the net horizontal
displacement of the projectile is $\mathbf{r} = \mathbf{v}_{horizontal}t_{flight}$

$= [(720\text{ m/s})(\cos 30°)\mathbf{i} + (12.5\text{ m/s})\mathbf{k}](73.5\text{ s}) = (45.8 \times 10^3\text{ m})\mathbf{i} + (918\text{ m})\mathbf{k}$. The total horizontal

range is $r = \sqrt{r_x^2 + r_z^2} = 45.8$ km. The change in the total range due to the ship's motion is
negligible to three significant digits; however, failure to compensate for the ship's forward speed
would cause the projectile to miss its target by nearly 1 km.

4-53. From the diagram

$v_{0x} = v_0 \cos \theta$

$v_{0y} = v_0 \sin \theta$ (i)

In the xy-frame

$x = v_{0x} t$ (ii)

$y = v_{0y} t - \dfrac{1}{2} g t^2$ (iii)

Also

$y = l \sin \alpha$

$x = l \cos \alpha$ (iv)

When the projectile hits the incline

$y = l \sin \alpha = v_{0y} t - \dfrac{1}{2} g t^2$ (v)

from (ii) $t = x/v_{0x} = l \cos \alpha / v_{0x}$ (vi)

Substitute t from (vi) in equation (v) to get the range as measured along the incline.

$$l = \frac{1}{\sin \alpha} \left\{ v_0 \sin \theta \cdot \frac{l \cos \alpha}{v_0 \cos \theta} - \frac{1}{2} g \frac{l^2 \cos^2 \alpha}{v_0^2 \cos^2 \theta} \right\}$$

Solving for l, gives

$$l = \frac{2 v_0^2}{g \cos \alpha} \left\{ \cos \theta \sin \theta - \tan \alpha \cos^2 \theta \right\}$$

$$\boxed{l = \frac{2 v_0^2 \cos^2 \theta}{g \cos \alpha} \left(\tan \theta - \tan \alpha \right)}$$

l will be maximum when $\dfrac{dl}{d\alpha} = 0$

$$\frac{dl}{d\theta} = \frac{2 v_0^2}{g \cos \alpha} \left(\cos 2\theta + (\tan \alpha) 2 \cos \theta \sin \theta \right) = 0$$

or $\cos 2\theta = - \tan \alpha \sin 2\theta$, $\cot 2\theta = - \tan \alpha = \cot(\alpha \pm \pi/2)$

Thus $\boxed{\theta = \dfrac{1}{2}(\alpha \pm \pi / 2)}$

4-55. The total instantaneous velocity of a point on the rim of the tire is the vector sum of the forward velocity of the tractor and the tangential velocity of the point, $\mathbf{v}_{tot} = \mathbf{v}_{tan} + \mathbf{u}$. This total instantaneous velocity must be zero when the point is on the road ($\theta = 270°$), or the tire will be slipping. Thus the tangential speed of a point on the rim must be equal to the forward speed u of the tractor.

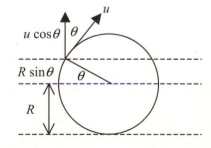

    In the reference frame of the tractor, the horizontal component of velocity of a glob is always zero and the vertical component of the velocity is $v_y = u \cos \theta$, which becomes the y component of the initial velocity v_{0y} of a launched glob. The maximum height above the launch point is

$$y_{max} = \frac{v_{0y}^2}{2g} = \frac{u^2 \cos^2 \theta}{2g}.$$ To find the maximum height above the ground, we must add the height

of the launch point above the ground, which is $R + R\sin\theta$ as shown in the diagram. So the height above the ground is $h = R(1 + \sin\theta) + \dfrac{u^2\cos^2\theta}{2g}$.

The maximum height depends on θ. To find the maximum possible height reached by a glob, differentiate h with respect to θ and set the derivative equal to zero:

$\dfrac{dh}{d\theta} = R\cos\theta - \dfrac{u^2\cos\theta\sin\theta}{g} = 0$, from which we get $\sin\theta_{max} = \dfrac{gR}{u^2}$. For R = 0.80 m and $u = 30$ km/h = 8.33 m/s, this gives $\theta_{max} = 6.49°$, from which we get

$h = (0.80\text{ m})(1 + \sin 6.49°) + \dfrac{(8.33\text{ m/s})^2(\cos 6.49°)^2}{2(9.81\text{ m/s}^2)} = \underline{4.4\text{ m}}.$

4-59. $a = g = \dfrac{v^2}{R} \Rightarrow v = \sqrt{gR} = \sqrt{(9.81\text{ m/s}^2)(200\text{ m})} = 44.3$ m/s. Use ω as defined in the solution to

4-58 to find the number of revolutions per minute required to give this acceleration: $\omega = \dfrac{v}{R} = \dfrac{44.3\text{ m/s}}{200\text{ m}} \times \dfrac{60\text{ s}}{\text{min}} \times \dfrac{1\text{ rev}}{2\pi\text{ radians}} = \underline{2.1\text{ rev/min}}.$

4-61. Let T stand for the time for one complete orbit. $v = \dfrac{2\pi R}{T} \Rightarrow a = \dfrac{v^2}{R} = \dfrac{4\pi^2 R}{T^2}$. For an orbit with a

radius of 6500 km and $T = 87$ min, this gives $a = \dfrac{4\pi^2(6500\times 10^3\text{ m})}{(87\text{ min})^2(60\text{ s/min})^2} = \underline{9.4\text{ m/s}^2}.$

4-63. $a = \dfrac{v^2}{r}$. Let f stand for the number of rotations per second. Then $v = 2\pi rf = 2\pi \times 0.1\text{ m} \times 1000/\text{s}$

$= 200\pi$ m/s. $a = \dfrac{v^2}{r} = \dfrac{(200\pi\text{ m/s})^2}{(0.10\text{ m})} = \underline{3.95\times 10^6\text{ m/s}^2}$. This is $= (3.95\times 10^6\text{ m/s}^2)/(9.81\text{ m/s}^2)\,g$

$= \underline{4.0\times 10^5\,g}.$

4-71. In the table shown, the acceleration was calculated using $a = \dfrac{4\pi^2 R}{T^2}$.

Planet	R (m)	T (yr)	a (m/s^2)	1/R^2 (m^{-2})
Mercury	5.79E+10	0.241	0.0395	2.98E-22
Venus	1.08E+11	0.615	0.0113	8.57E-23
Earth	1.50E+11	1.000	0.00595	4.44E-23

One way to see if the centripetal acceleration is inversely proportional to R^2 is to plot a graph of a vs $1/R^2$. The resulting plot is a straight line, proving the proportionality. (In the graph, the values of $1/R^2$ have been multiplied by 10^{22} to eliminate large negative exponents for the x axis values.)

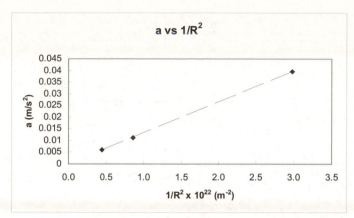

Another approach to the analysis is to calculate the log of a and R and plot a graph of log a vs log R. If a is inversely proportional to $1/R^2$, then this graph should be a straight line with a slope of -2.

log a	log R
−1.4034	10.76268
−1.94692	11.03342
−2.22548	11.17609

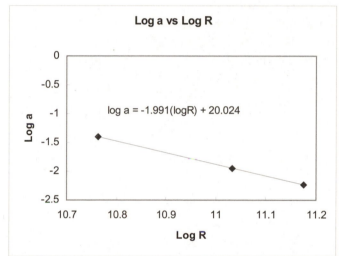

The graph was plotted in a spreadsheet, which was used to find the equation for the resulting line. The slope turns out to be almost exactly -2, again proving the desired proportionality.

4-79. $V = \left| \mathbf{v}_{train} + \mathbf{v}_{flea} \right|$, where $\mathbf{v}_{flea}$ is the velocity of the flea relative to the train. The cat is moving backwards relative to the train and the flea is moving backwards relative to the cat. The velocity of the flea relative to the train is 0.50 m/s − 0.10 m/s = 0.40 m/s backwards relative to the train. So the flea's speed relative to the ground is $V = \left| 5.00 \text{ m/s} - 0.40 \text{ m/s} \right| = \underline{4.60 \text{ m/s}}$.

4-87. $v = \sqrt{4.2^2 + 16^2 - 2(4.2)(16)\cos 70°}$
= 15.1 km/h
$x = 4.2 \sin 20° = 1.44$ km/h
$y = (16 - 1.44)$ km/h = 14.56 km/h
$\sin \theta = 14.56/v = 14.56/15.1$
$\theta = \sin^{-1}(0.96) = \underline{75°}$ (15° E of N)

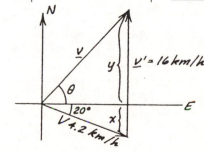

4-90. $\mathbf{v}_g = \mathbf{v} + \mathbf{V}$

(a) $v_g^2 = \sqrt{v^2 - v^2}$

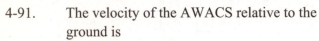

Total distance across
and back is $2d$.

$t = \text{dist}/v_g = 2d / \sqrt{v^2 - V^2}$

(b) $v_{up} = v - V$

$v_{down} = v + V$

$t_{up} = d/(v - V);\ t_{down} = d/(v + V)$

$t_{tot} = d/(v - V) + d/(v + V) = 2dv/(v^2 - V^2)$

The trip up and back takes longer (the denominator is larger)

4-91. The velocity of the AWACS relative to the
ground is
$\mathbf{v}_{AW} = 150\ \mathbf{i} + 750\ \mathbf{j}$ km/h
Relative to this, the UFO has velocity
$\mathbf{v}_{UFO} = -950 \cos 45°\ \mathbf{i} = 950 \sin 45°\ \mathbf{j}$ km/h $=$
$672\ \mathbf{i} - 672\ \mathbf{j}$ km/h
The velocity of the UFO relative
to the ground is
$\mathbf{v}_{g(UFO)} = \left(150\ \mathbf{i} + 750\ \mathbf{j}\right) + \left(-672\ \mathbf{i} - 672\ \mathbf{j}\right)$ km/h
$=\ 522\ \mathbf{i} + 78\ \mathbf{j}$ km/h
$v_{g(UFO)} = \sqrt{522^2 + 78^2} = 528$ km/h

The bearing is $\theta = \tan^{-1}\left(\dfrac{78}{522}\right) = 8.5°$ N of W

 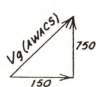

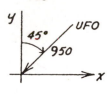

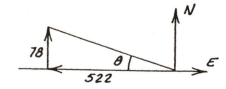

4-97. (a) $\mathbf{r} = (6.0 + 2.0t^2)\mathbf{i} + (3.0 - 2.0t + 3.0t^2)\mathbf{j}$. $\mathbf{v} = \dfrac{d\mathbf{r}}{dt} = 4.0t\mathbf{i} + (-2.0 + 6.0t)\mathbf{j}$. At $t = 2.0$ s,

$\mathbf{v} = 8.0\mathbf{i} + 10\mathbf{j}$ m/s, so the speed is $v = \sqrt{v_x^2 + v_y^2} = \sqrt{(8.0\ \text{m/s})^2 + (10\ \text{m/s})^2} = 13$ m/s.

(b) $\mathbf{a} = \dfrac{d\mathbf{v}}{dt} = 4.0\mathbf{i} + 6.0\mathbf{j}$ m/s² at all times. The magnitude of the acceleration is

$a = \sqrt{a_x^2 + a_y^2} = \sqrt{(4.0\ \text{m/s}^2)^2 + (6.0\ \text{m/s}^2)^2} = 7.2$ m/s². The direction relative to the x axis is

$\theta = \tan^{-1}\dfrac{v_y}{v_x} = \tan^{-1}\dfrac{6.0}{4.0} = 56.3°.$

4-101. $y = y_0 + v_{0y}t - \dfrac{gt^2}{2}$, $x = v_{0x}t$. When $y = 0$, $x = x_{max}$. For $\theta = 45°$, $v_{0x} = v_{0y} = v_0/\sqrt{2}$. Then the

time to reach $y = 0$ is $t = \dfrac{\left(\dfrac{v_0}{\sqrt{2}}\right) + \sqrt{\dfrac{v_0^2}{2} + 2y_0g}}{g} = \dfrac{v_0 + \sqrt{v_0^2 + 4y_0g}}{g\sqrt{2}}$. (There is another value of t

that solves the equation for $y = 0$, but it will be negative because the square root in the quadratic formula will be larger than v_0. Choose the positive value.) The horizontal range is

$x_{max} = \dfrac{v_0}{\sqrt{2}}\left(\dfrac{v_0 + \sqrt{v_0^2 + 4y_0g}}{g\sqrt{2}}\right) = \dfrac{v_0^2 + v_0\sqrt{v_0^2 + 4y_0g}}{2g}$. Rearrange: $2gx_{max} - v_0^2 = v_0\sqrt{v_0^2 + 4y_0g}$.

Square both sides: $4g^2x_{max}^2 - 4gx_{max}v_0^2 + v_0^4 = v_0^4 + 4gy_0v_0^2$. Cancel the fourth degree terms and the remaining common factor of 4 and solve for v_0:

$v_0 = \sqrt{\dfrac{x_{max}^2 g}{x_{max} + y_0}} = \sqrt{\dfrac{(70.87\ \text{m})^2(9.81\ \text{m/s}^2)}{70.87\ \text{m} + 2.0\ \text{m}}} =$

<u>26 m/s</u>.

4-105. (a) $V = |v_{car} - v_{truck}| = |90\ \text{km/h} - 60\ \text{km/h}| = 30\ \text{km/h} = 8.33\ \text{m/s}$.

(b) In the reference frame of the truck, the car has to travel a total distance of 90 m: It must travel 40 m to catch up to the truck, 10 m to pass the truck, and then travel another 40 m to get ahead of the truck. The time for the car to travel 90 m at 8.33 m/s is $t = \dfrac{90\ \text{m}}{8.33\ \text{m/s}} = \underline{10.8\ \text{s}}$.

5-5. $\bar{a} = \dfrac{v_2 - v_1}{\Delta t} = \dfrac{(80 \text{ km/h} - 0 \text{ km/h})}{5.8 \text{ s}} \times \dfrac{1000}{\text{km}} \dfrac{\text{m}}{} \times \dfrac{1\text{h}}{3600 \text{ s}} = 3.83 \text{ m/s}^2.$ Final answer

$= \underline{3.8 \text{ m/s}^2}.$ To find the magnitude of the average force, use the three-digit intermediate result for

$\bar{a}: \bar{F} = m\bar{a} = 1620 \text{ kg} \times 3.83 \dfrac{\text{m}}{\text{s}^2} = 6205 \text{ N}.$ Final answer $= \underline{6.2 \times 10^3 \text{ N}}.$

5-9.   Vector note: "Decelerates" is a nontechnical way of
stating that the acceleration points in the opposite
direction from the velocity, causing the speed to
decrease. If we take the positive direction for vectors to
be in the direction of motion, then a must be represented
by a negative number, and the net force will also be a
negative number. A "free-body" diagram is shown to
illustrate these concepts. Thus

Direction of motion

$F_{net} = ma = 1500 \text{ kg} \times (-8.0 \text{ m/s}^2) = \underline{-1.2 \times 10^3 \text{ N}}.$

Again, the – sign means the force points in the opposite
direction from the original motion.

5-15. $\bar{a} = \dfrac{\Delta v}{\Delta t}.$ First interval: $\bar{a}_1 = \dfrac{10.0 \text{ m/s} - 15.0 \text{ m/s}}{1.2 \text{ s}} = -4.17 \text{ m/s}^2.$ Final result $= \underline{-4.2\dfrac{\text{m}}{\text{s}^2}}.$ Second

interval: $\bar{a}_2 = \dfrac{5.0 \text{ m/s} - 10.0 \text{ m/s}}{2.1 \text{ s}} = -2.38 \text{ m/s}^2.$ Final result $= \underline{-2.4\text{m/s}^2}.$ During the first

interval, $\bar{F}_1 = m\bar{a}_1 = 240 \text{ kg} \times (-4.17 \text{ m/s}^2) = \underline{-1.0 \times 10^3 \text{N}}.$ During the second interval,

$\bar{F}_2 = m\bar{a}_2 = 240 \text{ kg} \times (-2.38 \text{ m/s}^2) = \underline{-5.7 \times 10^2 \text{N}}.$ Vector note: The – signs imply that the

accelerations and forces point in the opposite direction from the motion.

5-19.   For the first 0.30-s interval, $\bar{a} = \dfrac{\Delta v}{\Delta t} = \dfrac{638 \text{ m/s} - 657 \text{ m/s}}{0.30 \text{ s}} = -63.3 \text{ m/s}^2.$ The mass is

$m = 100 \text{ lb} \times \dfrac{1 \text{ kg}}{2.205 \text{ lb}} = 45.4 \text{ kg}.$ $\bar{F} = m\bar{a} = 45.4 \text{ kg} \times (-63.3 \text{ m/s}^2) = \underline{-2.9 \times 10^3 \text{ N}}.$

For the last 0.30 s interval, $\bar{a} = \dfrac{502 \text{ m/s} - 514 \text{ m/s}}{0.30 \text{ s}} = \underline{-40.0 \text{ m/s}^2}.$

$\bar{F} = m\bar{a} = 45.4 \text{ kg} \times (-40.0 \text{ m/s}^2) = \underline{-1.8 \times 10^3 \text{ N}}.$ Vector note: The – signs imply that the

accelerations and forces point in the opposite direction from the motion.

5-21. $v = \dfrac{dx}{dt} = \dfrac{d}{dt}x_0\left[1 - \cos(bt)\right] = 0 + bx_0 \sin(bt).$ $\underline{v = bx_0 \sin(bt)}.$

$a = \dfrac{dv}{dt} = \dfrac{d}{dt}bx_0 \sin(bt) = b^2 x_0 \cos(bt).$ $\underline{F = ma = mb^2 x_0 \cos(bt)}.$

$x = x_0\left[1 - \cos(bt)\right] \Rightarrow x_0 \cos(bt) = -(x - x_0).$ $\underline{F = -mb^2(x - x_0)}.$ This type of motion, in

which the magnitude of the force is proportional to the distance from an equilibrium point and
always points in the opposite direction from the displacement, turns out to be very important in
science.

5-25.   Find the east and south components of **P** that will make the net force zero:

$$F_{not,E} = \sum F_E = P_E - (240 \text{ N}) \cos 30° = 0 \Rightarrow P_E = 208 \text{ N}.$$

$$F_{net,N} = \sum F_N = 270 \text{ N} - P_S - (240 \text{ N}) \sin 30° = 0 \Rightarrow P_S = 150 \text{ N}.$$

$$|\mathbf{P}| = \sqrt{P_E^2 + P_S^2} = \sqrt{(208 \text{ N})^2 + (150 \text{ N})^2} = \underline{256 \text{ N}}.$$

$$\theta = \tan^{-1} \frac{P_S}{P_E} = \tan^{-1} \frac{208 \text{ N}}{150 \text{ N}} = \underline{54.2°} \quad (54.2° \text{ S of E}).$$

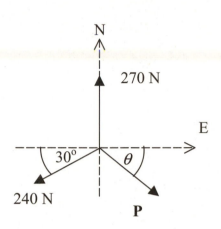

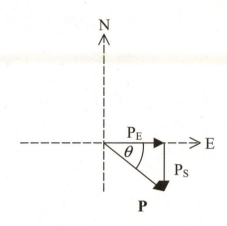

5-27.   The net force in the east direction is $F_{net,E} = (2500 \text{ N}) \sin 15° + (3200 \text{ N}) \sin 30° = 2247 \text{ N}$. The net force in the north direction is $F_{net,N} = (2500 \text{ N}) \cos 15° + (3200 \text{ N}) \cos 30° = 5186 \text{ N}$.

To find the east and north components of acceleration, divide each force component by the mass:

$$a_E = \frac{F_{net,E}}{m} = \frac{2247 \text{ N}}{1400 \text{ kg}} = 1.605 \text{ m/s}^2 \qquad a_N = \frac{F_{net,N}}{m} = \frac{5186 \text{ N}}{1400 \text{ kg}} = 3.704 \text{ m/s}^2$$

$$a = \sqrt{a_E^2 + a_N^2} = \sqrt{\left(1.605 \text{ m/s}^2\right)^2 + \left(3.704 \text{ m/s}^2\right)^2} = \underline{4.037 \text{ m/s}^2}.$$

$$\theta = \tan^{-1} \frac{a_E}{a_N} = \tan^{-1} \frac{1.605}{3.704} = \underline{23.43°}.$$

This is 23.43° E of N.

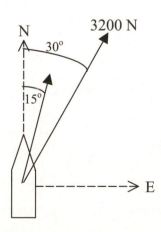

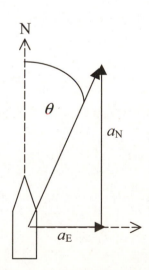

5-31. Define x and y axes with the x direction perpendicular to the dock and the y direction along the dock. Resolve the two 360-N forces into x and y components. By symmetry the y components cancel. The resultant force in the x direction is

$R_x = 2 \times 260 \quad \text{N} + 2 \times (360 \text{ N}) \sin 20° = 766 \text{ N}.$

The resultant expressed as a vector is

R = 766**i** + 0**j** N.

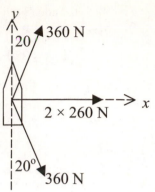

5-35. $m = \dfrac{W_{Earth}}{g_{Earth}} = \dfrac{750 \quad \text{N}}{9.81 \text{ m/s}^2} = \underline{76.5 \text{ kg}}$. Mass is an intrinsic property of the object, so its mass is

76.5 kg on Earth, Mars, and Jupiter. Its weight depends on the value of g. The calculations can be simplified using proportional reasoning:

$W_{Mars} = mg_{Mars} = W_{Earth} \dfrac{g_{Mars}}{g_{Earth}} = 0.38(750 \text{ N}) = 285 \text{ N}$

$W_{Jupiter} = mg_{Jupiter} = W_{Earth} \dfrac{g_{Jupiter}}{g_{Earth}} = 2.53(750 \text{ N}) = 1.90 \times 10^3 \text{ N}$

5-39. Draw two "free–body" diagrams:

Bottom chandelier: Top chandelier:

   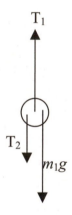

$T_2 - m_2 g = 0$ $T_1 - T_2 - m_1 g = 0$

$T_2 = m_2 g = \underline{29.4 \text{ N}}$ $T_1 = T_2 + m_1 g = m_2 g + m_1 g = \underline{128 \text{ N}}$

5-41. "Free-body" diagrams:

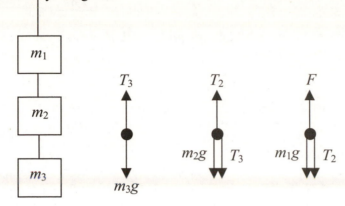

The system is described as being stationary, so it must be in equilibrium. Thus the net force on each mass must be zero. For m_3, $T_3 - m_3 g = 0 \Rightarrow T_3 = m_3 g$. For m_2,

$T_2 - T_3 - m_2 g = 0 \Rightarrow T_2 = T_3 + m_2 g = (m_2 + m_3)g$. For m_1, $F - T_2 - m_1 g = 0 \Rightarrow$

$F = T_2 + m_1 g = (m_1 + m_2 + m_3)g$.

5-43. Suppose another sailor pulls on the free end of the rope with a force P as shown. Then the "free-body" diagram for the sailor in the seat has two forces, and the net force on the sailor is $F_{net} = P - Mg$. The reaction from the rope acts on the puller. For the sailor to move up, P must at least be equal to Mg. If the sailor in the seat reaches over and pulls on the rope, the reaction from the rope pulls up on the sailor in the seat. Now the "free-body" diagram shows *two* upward forces on the sailor in the seat, and the net force is $F_{net} = 2P - Mg$. Now the minimum force required for the sailor to lift himself is $Mg/2$.

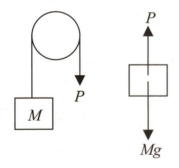

Another sailor pulls on the rope.

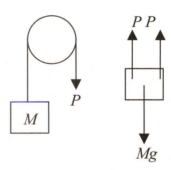

The sailor in the seat pulls himself up.

5-47. The 60-N force acts on a total mass of 50 kg, so $a = \dfrac{F}{m_{total}} = \dfrac{60 \text{ N}}{50 \text{ kg}} = \underline{1.2 \text{ m/s}^2}$. To find the force each mass exerts on the other, use these "free-body" diagrams with R as the unknown force, assuming the 60-N force points to the right:

The vertical forces aren't shown because they add to zero. Applying Newton's Second Law to the 30-kg mass gives $R = 30a = \underline{36 \text{ N}}$. The 20-kg box exerts a 36-N force to the right on the 30-kg box, and the 30-kg box exerts a 36-N force to the left on the 20 kg box. (Note: To check, note that the net force on the 20 kg box is $(60 - 36) \text{ N} = 24 \text{ N}$. This gives an acceleration of 1.2 m/s².)

5-53. The net force is the vector sum of the force exerted by the wind plus the tension in the string. (The weight of the balloon is assumed to be much less than the other forces acting and can be neglected.) Take the +y direction to point up and the +x direction to point right. Then

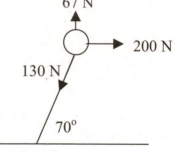

$F_{net,x} = 200 \text{ N} - (130 \text{ N}) \cos 70° = 156 \text{ N}$

$F_{net,y} = 67 \text{ N} - (130 \text{ N}) \sin 70° = -55.2 \text{ N}$

$|\mathbf{F}_{net}| = \sqrt{R_x^2 + R_y^2} = \underline{165 \text{ N}}$

$\theta = \tan^{-1} \dfrac{R_y}{R_x} = \underline{-19.4°}$ (19.4° *below* horizontal, or +x direction)

(A dynamics note: because the net force is not zero, the balloon is not in equilibrium at the instant described in this problem.)

5-59. A diagram and three "free-body" diagrams are shown. Write Newton's Second Law for each of the three masses:

$F - m_1 g - T_2 = m_1 a_1$

$T_2 - m_2 g - T_3 = m_2 a_2$

$T_3 - m_3 g_2 = m_3 a_3$

All three masses must have the same acceleration: $a_1 = a_2 = a_3$, which we'll just call a. To find a, add all three equations together. Then T_2 and T_3 cancel, giving

$F - (m_1 + m_2 + m_3) g = (m_1 + m_2 + m_3) a$

$\Rightarrow a = \dfrac{F - (m_1 + m_2 + m_3) g}{m_1 + m_2 + m_3}$.

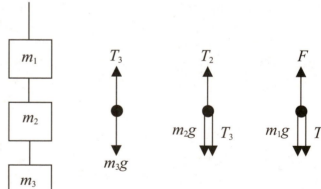

Then the tension in the first cable is $\underline{F}$, and the tensions in the second and third cables are

$$T_1 = F - m_1(g + a) = F\left(\frac{m_2 + m_3}{m_1 + m_2 + m_3}\right)$$ and

$$T_2 = m_3(g + a) = F\left(\frac{m_3}{m_1 + m_2 + m_3}\right).$$

5-61. The reading on the scale is the normal force exerted by the scale. The net force on the passenger is $N - mg = ma \Rightarrow N = mg + ma$, choosing up to be the positive direction for vectors. The man's mass is 220 lbm = 99.8 kg, and his weight is $mg = 220$ lbf = 979 N. If $a = 1.6$ m/s^2, then $N = (979\text{ N} + 99.8\text{ kg} \times 1.6\text{ m/s}^2) = \underline{1139\text{ N} = 256\text{ lb}}$. If the elevator has a downward acceleration, then $a = -1.6$ m/s^2, and $\underline{N = 819}$ $\underline{\text{N} = 184\text{ lb}}$.

5-63. A "free-body" diagram is shown for one of the dice. From the diagram,
$T \cos\theta - mg = 0$, $T \sin\theta = ma$
where the positive horizontal direction is assumed to point to the right, and the positive vertical direction is up. Dividing the bottom equation by the top equation gives $\tan\theta = \dfrac{a}{g}$, so

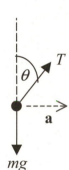

$$\theta = \tan^{-1}\left(\frac{2.5\text{ m/s}^2}{9.81\text{ m/s}^2}\right) = 14.3°.\text{ The final result} = \underline{14°}.\text{ The tension in}$$

the string for each die is $T = \dfrac{mg}{\cos\theta} = \dfrac{0.025\text{ kg} \times 9.81\text{ m/s}^2}{\cos 14.3°} =$ $\underline{0.25\text{ N}}$.

5-81. $\mathbf{F}_{total} = \mathbf{F}_1 + \mathbf{F}_2 = (2 - 4)\mathbf{i} + (-5 + 8)\mathbf{j} + (3 + 1)\mathbf{k} = \underline{-4\mathbf{i} + 3\mathbf{j} + 4\mathbf{k}\text{ N}}.$

$$\mathbf{a} = \frac{\mathbf{F}_{total}}{m} = \frac{-4\mathbf{i} + 3\mathbf{j} + 4\mathbf{k}\text{ N}}{6.0\text{ kg}} = -\frac{2}{3}\mathbf{i} + \frac{1}{2}\mathbf{j} + \frac{2}{3}\mathbf{k}\text{ m/s}^2 = \underline{-0.67\mathbf{i} + 0.50\mathbf{j} + 0.67\mathbf{k}\quad\text{m/s}^2}.$$

$$|\mathbf{a}| = \sqrt{a_x^2 + a_y^2 + a_z^2} = \sqrt{\left(\frac{2}{3}\right)^2 + \left(\frac{1}{2}\right)^2 + \left(\frac{2}{3}\right)^2} = \underline{1.1\text{ m/s}^2}.$$

5-83. (a) Represent the boat as a point mass. The water exerts an upward force with magnitude B on the boat. The tension T is shown resolved into x and y components. The angle θ is found from the fact that the rope is 50 m long and the depth of the water is 10 m: $\theta = \sin^{-1}\left(\dfrac{10\text{ m}}{50\text{ m}}\right) = 11.5°$.

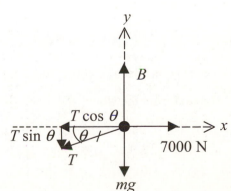

(b) The boat is in equilibrium so the sum of all the forces on the boat must be zero. The sum of x components will give the tension T:

$$7000\ \text{N} - T\cos\theta = 0 \Rightarrow T = \frac{7000\ \text{N}}{\cos 11.5^\circ} = \underline{7.14 \times 10^3\ \text{N}}.$$

(c) To find B, use the sum of the y components: $B - mg - T\sin\theta = 0$

$$\Rightarrow B = mg + T\sin\theta = (2500\ \text{kg})(9.81\ \text{m/s}^2) + (7.14 \times 10^3\ \text{N})(\sin 11.5^\circ) = \underline{2.59 \times 10^4\ \text{N}}.$$

Vector note: $\mathbf{B} = (2.59 \times 10^4\ \text{N})\mathbf{j}$.

5-85.   (a) Represent each child by a block, and represent each end of the rope by points between them. Only the horizontal forces are shown. The vertical weight of each child and the vertical normal force exerted by the ground on each child have been omitted from the diagrams.

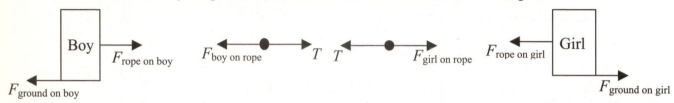

(b) By Newton's Third Law, $\left|F_{\text{rope on boy}}\right| = \left|F_{\text{boy on rope}}\right|$. Because $F_{\text{boy on rope}} = 250$ N, the magnitude of $F_{\text{rope on boy}} = 250$ N. The boy doesn't move, so the net horizontal force on him must be zero, and that means the magnitude of $F_{\text{ground on boy}} = \underline{250\ \text{N}}$. By similar reasoning, the magnitude of $F_{\text{ground on girl}} = \underline{250\ \text{N}}$.

(c) The net horizontal force on the rope must be zero, so $T = F_{\text{boy on rope}} = \underline{250\ \text{N}}$.

(d) Nothing changes except the origin of the force on the girl's end of the rope. Suppose the rope passed through a hole in a wall so the boy couldn't see what was at the other end. He would not be able to tell whether there was a person at the other end or if the end was tied to something immovable.

5-87.

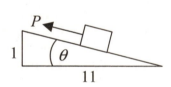

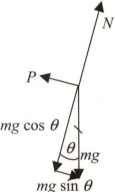

(a) A sketch is shown, and the "free-body" diagram is shown on the right. The weight of the boxcar has been resolved into components along the incline and normal to it. The angle θ can be found from the slope: $\theta = \tan^{-1}\dfrac{1}{11} = 5.19^\circ$.

(b) For the boxcar to move at constant speed, its acceleration along the incline must be zero. Thus the sum of forces along the incline must be zero: $P - mg\sin\theta = 0$, from which we get

$$P = mg\sin\theta = (20 \times 10^3\ \text{kg})(9.81\ \text{m/s}^2)(\sin 5.19^\circ) = \underline{1.8 \times 10^4\ \text{N}}.$$ Vector note: As shown in the diagram, $\mathbf{P} = 1.8 \times 10^4$ N up the incline.

(c) In the absence of any friction forces, it makes no difference whether the car is moving at constant velocity up the incline or down the incline. The force required is 1.8×10^4 N up the incline in either case.

5-89.   The scale reading is equal to the normal force exerted on the woman by the scale. Newton's Second Law says $\sum F = N - mg = ma$, where m is the woman's mass (60 kg). The scale reading is $N = m(g + a)$. Choose the positive direction to be up.

(a) At rest $\Rightarrow a = 0 \Rightarrow N = mg = (60\text{ kg})(9.81\text{ m/s}^2) = \underline{589\text{ N}}$

(b) Accelerating up $\Rightarrow N = (60\text{ kg})(9.81\text{ m/s}^2 + 1.8\text{ m/s}^2) = \underline{697\text{ N.}}$

(c) Constant velocity $\Rightarrow a = 0 \Rightarrow N = mg = (60\text{ kg})(9.81\text{ m/s}^2) = \underline{589\text{ N.}}$

(d) Free fall $\Rightarrow a = -g \Rightarrow N = \underline{0\text{ N.}}$

5-91.   (a) A sketch and "free-body" diagram are shown with the initial velocity of the car shown in the sketch. Choose the $+x$ direction to point leftward up the incline in the direction of the original motion. The weight of the car has its x component pointing in the $-x$ direction. The sum of forces normal to the incline is zero, and there is only one force along the incline:

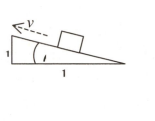

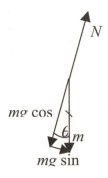

$F_{net,x} = \sum F_x = -mg\sin\theta = ma_x$. The angle θ is found from the slope: $\theta = \tan^{-1}\dfrac{1}{10} = 5.71°$.

The acceleration is $a_x = -g\sin\theta = -(9.81\text{ m/s}^2)\sin 5.71° = \underline{-0.976\text{ m/s}^2}$. The – sign means the acceleration points down the incline, opposite the direction of the car's initial velocity. In non–technical language, this is the car's "deceleration."

(b) We can use $v^2 = v_0^2 + 2a_x x$, where $v = 0$ and $v_0 = 50\dfrac{\text{km}}{\text{h}} = 13.9\dfrac{\text{m}}{\text{s}}$. Solving for x gives

$$x = -\frac{v_0^2}{2a_x} = -\frac{(13.9\text{ m/s})^2}{2(-0.976\text{ m/s}^2)} = \underline{99.0\text{ m.}}$$

(c) Even when the car momentarily comes to rest and then begins to roll back down the incline, it continues to have an acceleration $a_x = -0.976\dfrac{\text{m}}{\text{s}^2}$. To find its speed when it gets back to its starting point, we can again use $v^2 = v_0^2 + 2a_x x$, but caution must be used in dealing with the vector quantities in the equation. Now $v_0 = 0$, and both a_x and x point *down* the incline and must be entered as negative quantities in the equation. We get

$v = \sqrt{2a_x x} = \sqrt{2(-0.976\text{ m/s}^2)(-99.0\ m)} = \underline{13.9\text{ m/s.}}$ (This is the car's *speed*. Its *velocity* is $-(13.9\text{ m/s})\mathbf{i}$ because it is now moving *down* the incline.)

5-93.

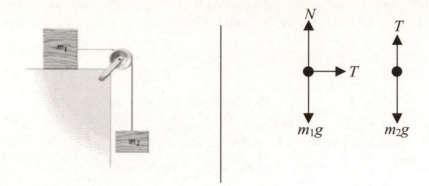

(a) Take the x direction to point right and the y direction to point up. Newton's Second Law for m_1 gives $\sum F_y = N - m_1 g = 0$, $\sum F_x = T = m_1 a_1$. For m_2, $\sum F_y = T - m_2 g = m_2 a_2$. If the string passing over the pulley does not stretch and has negligible mass, then the magnitude of a_2 must be the same as the magnitude of a_1. If m_1 moves in the $+x$ direction, then m_2 must move in the $-y$ direction, which means $a_2 = -a_1$. Then the equations for $\sum F_x$ for m_1 and $\sum F_y$ for m_2 can be combined to give an equation for a_1: $a_1 = \dfrac{m_2 g}{m_1 + m_2}$.

(b) The equation for $\sum F_x$ gives the tension: $T = m_1 a_1 = \dfrac{m_1 m_2 g}{m_1 + m_2}$.

6-1.   Pull $= nP_{Egyption}$, where $n =$ number of Egyptians. Assume that the obelisk is being pulled at a constant speed. The total pull must be equal to the kinetic friction force acting on the obelisk: Pull $= f = \mu_k N$. For an object resting on a horizontal surface (flat ground) with no vertical forces other than gravity and the normal force acting, the normal force N is equal to the object's weight mg. Thus $nP_{Egyption} = \mu_k mg \Rightarrow n = \dfrac{\mu_k mg}{P_{Egyption}} = \dfrac{(0.30)(7000 \text{ kg})(9.81 \text{ m/s}^2)}{360 \text{ N}} = \underline{5.7 \times 10^3}$

Egyptians.

6-5.   Assume that that the car has antilock brakes so it stops without sliding. Then the force between the tire and the road is always the force of static friction. The minimum stopping distance will occur for the largest friction force, which is $f_{s,\max} = \mu_s N$. On a level surface with no vertical forces acting other than gravity, $N = mg \Rightarrow f_{s,\max} = \mu_s mg$. The friction force is opposite the motion, so $ma_{dry} = -f_{s,\max} = -\mu_{s,dry} mg \Rightarrow a_{dry} = -\mu_{s,dry} g$. $a_{dry} = \dfrac{v^2 - v_0^2}{2x_{dry}} = -\dfrac{v_0^2}{2x_{dry}}$, which

means $\dfrac{v_0^2}{2x_{dry}} = \mu_{s,dry} g$. For the icy road, $\dfrac{v_0^2}{2x_{icy}} = \mu_{s,icy} g$. Combining the two equations gives

$\dfrac{x_{icy}}{x_{dry}} = \dfrac{\mu_{s,dry}}{\mu_{s,icy}}$, from which we get $x_{icy} = x_{dry} \dfrac{\mu_{s,dry}}{\mu_{s,icy}} = (38 \text{ m})\left(\dfrac{0.85}{0.20}\right) = \underline{1.6 \times 10^2 \text{ m}}$.

6-7.   The technique is similar to that outlined in Problem 6-5 except that we use the force of kinetic friction, which means $a = -\mu_k g$. $2ax = v^2 - v_0^2 \Rightarrow x = -\dfrac{v_0^2}{2a} = \dfrac{v_0^2}{2\mu_k g}$. $v_0 = 10$ km/h $= 25$ m/s

$\Rightarrow x = \dfrac{(25 \text{ m/s})^2}{2(0.60)(9.81 \text{ m/s}^2)} = \underline{53 \text{ m}}$.

6-9. $f_k = \mu_k N \Rightarrow a = -\dfrac{f_k}{m} = -\dfrac{\mu mg}{m} = -\mu g$, where the $-$ sign means the acceleration points in the opposite direction from the velocity. Since the player is on level ground with no vertical forces other than gravity and the normal force acting, $N = mg$. $\mu k = 0.30 \Rightarrow a = -(0.30)(9.81 \text{ m/s}^2) = -2.94$ m/s^2. The distance required for him to stop is $x = \dfrac{v^2 - v_0^2}{2a} = \dfrac{0 - (4.5 \text{ m/s})^2}{2(-2.94 \text{ m/s}^2)} = 3.4$ m. Since he only has to slide 2.8 m to reach home, he will easily reach the plate. His speed when he reaches the plate will be $v = \sqrt{v_0^2 + 2ax} = \sqrt{(4.5 \text{ m/s})^2 + 2(-2.94 \text{ m/s}^2)(2.8 \text{ m})} = \underline{1.9 \text{ m/s}}$.

6-13.   The net horizontal force acting on the car is $f_{road} - f_{air} = ma \Rightarrow f_{road} = ma + f_{air} = ma + \dfrac{C\rho A v^2}{2}$.

Substituting the numerical data from the problem,

$f_{air} = (900 \text{ kg})(2.0 \text{ m/s}^2) + \dfrac{(0.30)(1.3 \text{ kg/m}^3)(2.8 \text{ m}^2)(25 \text{ m/s})^2}{2} = \underline{2.1 \times 10^3 \text{ N}}$.

6-17.   A free body diagram for the sled is shown. P is the pull exerted
by the girl, m is the mass of the sled, N is the normal force
exerted by the ground on the sled, and f is the kinetic friction
force between the sled and the ground. Take x to point right
and y to point up. Then $\sum F_y = N + P\sin 30° - mg = 0$

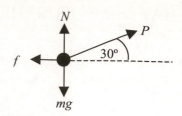

because the sled does not move in the y direction.

$\sum F_x = P\cos 30° - f = 0$ because the sled is being pulled at constant velocity. From the
equation for $\sum F_y$, $N = mg - P\sin 30°$. From the equation for

$\sum F_x$, $P\cos 30° - f = P\cos 30° - \mu_k N = P\cos 30° - \mu_k N = 0$. Thus

$$P\cos 30° - \mu_k(mg - P\sin 30°) = 0, \text{ which gives } P = \frac{\mu_k mg}{\cos 30° + \mu_k \sin 30°}$$

$$= \frac{(0.60)(40 \text{ kg})(9.81 \text{ m/s}^2)}{\cos 30° + (0.60)\sin 30°} = 2.0 \times 10^2 \text{ N}.$$

6-19.   "Deceleration" is a nontechnical way of saying that the speed
of the truck is decreasing. That means the magnitude of the
velocity is decreasing. Suppose the truck's initial velocity
was to the right $(+x)$ as shown. If it is slowing down, then its
acceleration must point to the left $(-x)$. If the box doesn't
slide on the back of the truck, then it must always be at rest
relative to the truck and must have the same acceleration as
the truck. If the box is sliding, that means its acceleration is
different from the truck's. The free body diagram for the box
shows that there are only three forces acting on it: its weight
mg, a normal force N, and a friction force f between it and the
truck. To determine the direction of the friction force,
suppose there was no friction between the box and the truck.

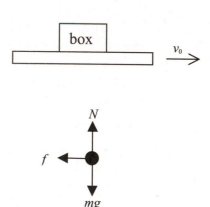

Then when the truck began to slow down, the box would continue moving to the right at its
original speed. That means the friction force must point to the left as shown in the diagram. This
is very important: *There is no horizontal force in the direction of motion*!

The vertical forces add to zero, so $N = mg$. The box is sliding, so f is a force of kinetic friction.
Since f is the only horizontal force acting, $-f = -\mu_k N = ma_x$. Since $N = mg$, this gives
$a_x = -\mu_k g = -(0.50)(9.81 \text{ m/s}^2) = -4.91 \text{ m/s}^2$. The acceleration of the box relative to the truck
is $\mathbf{a}' = \mathbf{a}_x - \mathbf{a}_{truck}$, which gives $a' = -4.91 \text{ m/s}^2 - (-7.0 \text{ m/s}^2) = 2.01 \text{ m/s}^2$. (The + sign means the
box is moving to the right relative to the truck.) If the box slides a distance x along the back of the
truck, then its speed relative to the truck is given by $(v')^2 - (v')_0^2 = 2a'x$, where $v_0' = 0$ because
the box was initially at rest relative to the truck (it was moving with the same velocity as the
truck). Thus $v' = \sqrt{2ax} = \sqrt{2(2.01 \text{ m/s}^2)(2.0 \text{ m})} = 2.8 \text{ m/s}$.

6-23. A free body diagram is shown. As the plate is tilted, the block remains at rest as the static friction force increases from zero (at zero elevation) to its maximum value of $\mu_s N$ at some angle θ, and then the block begins to slide. At θ, since the block is at rest, the sums of forces along the plate and perpendicular to the plate are

$$N = mg\cos\theta$$

$$\mu_s N = mg\sin\theta$$

Dividing the second equation by the first gives

$$\mu_s = \frac{\sin\theta}{\cos\theta} = \tan\theta. \text{ For } \theta = 38°, \ \underline{\mu_s = 0.78}.$$

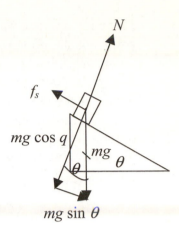

$mg\cos q$

$mg\sin\theta$

6-25. At terminal speed, $a = 0$, which means viscous drag force = weight $\Rightarrow bv_T = mg$.

$$v_T = \frac{mg}{b} = \frac{(3.9\times 10^{-9}\text{ kg})(9.81\text{ m/s}^2)}{2.8\times 10^{-5}\text{ kg/s}} = \underline{1.4\times 10^{-3}\text{ m/s (1.4 mm/s)}}.$$

6-27. Recall that $f_k = \mu_k N$. On a level road with no vertical forces acting except gravity, $N = mg$, so $a_{level} = -\mu_k g$ if friction is the only horizontal force acting. On a slope of angle θ, $N = mg\cos\theta$. If the car is initially moving down the slope, then there is a component of the car's weight pointing down the slope and the friction force points up the slope. Taking the + direction to be downward (in the direction of the initial velocity), then the acceleration along the slope is $a_{slope} = g(\sin\theta - \mu_k\cos\theta)$. If the car is going to slow down, the acceleration must be negative (point in the opposite direction from the initial velocity). For the slope, this means $\mu_k\cos\theta > \sin\theta$, or the car won't be able to slide to a stop. For both cases, use $v^2 = v_0^2 + 2ax$ with $v_0 = 90$ kph = 25 m/s and $v = 0$. Assume μ_k is the same on the level road and the slope. Solve for a on the level road and use that to find μ_k, then find a and x on the slope. Level road:

$$a_{level} = -\frac{v_0^2}{2x_{level}} = -\frac{(25\text{ m/s})^2}{2(35\text{ m})} = -8.93\text{ m/s}^2. \ \mu_k = -\frac{a_{level}}{g} = \frac{8.93}{9.81} = 0.910. \text{ For a 1:10 slope,}$$

$$\theta = \tan^{-1}\frac{1}{10} = 5.71°. \ a_{slope} = (9.81\text{ m/s}^2)[\sin 5.71° - (0.910)\cos 5.71°] = -7.91\text{ m/s}^2.$$

$$x_{slope} = -\frac{v_0^2}{2a_{slope}} = \frac{(25\text{ m/s})^2}{2(7.91\text{ m/s}^2)} = \underline{39.5\text{ m}}.$$

6-31. In Example 3, change the angle from 30° to some arbitrary value θ. Then requirement for the minimum magnitude of P is $P\cos\theta - \mu_k P\sin\theta - \mu_k mg = 0$, or $P(\cos\theta - \mu_k\sin\theta) - \mu_k mg = 0$. If $\cos\theta - \mu_k\sin\theta < 0$, then both terms in the equation will be negative and it will not be possible to find a value for P that satisfies the equation. If $\mu_k\sin\theta - \cos\theta > 0$, the crate will not move no matter how hard it is pushed. This can be rewritten as $\mu_k\sin\theta > \cos\theta$, or $\tan\theta > \dfrac{1}{\mu_k}$, or

$$\underline{\theta > \tan^{-1}\left(\frac{1}{\mu_k}\right)}.$$

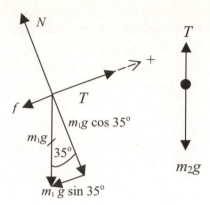

6-33.   Free body diagrams for the two masses are shown. Let the tension on the string be T. Take the $+x$ direction for m_1 to point up the ramp. For m_2, take the $+$ direction to point up. The sum of forces perpendicular to the ramp is zero because there's no motion in that direction. Thus $N = m_1 g \cos 35°$.

The magnitude of the kinetic friction force is $\mu_k N$, so the sum of forces along the x direction is $T - \mu_k m_1 g \cos 35° - m_1 g \sin 35° = m_1 a_1$. For m_2, the sum of forces gives $T - m_2 g = m_2 a_2$. If m_2 moves down, then m_1 must move up the ramp. Thus the accelerations a_1 and a_2 must have opposite signs. If the string connecting them doesn't stretch and has negligible mass, then the accelerations must have the same magnitude. Thus we conclude that $a_2 = -a_1$, which we'll just call a. Then the two force equations become $T - m_1 g(\sin 35° + \mu_k \cos 35°) = m_1 a$ and $m_2 g - T = m_2 a$. Adding the two equations eliminates T and gives $a = \dfrac{g[m_2 - m_1(\sin 35° + \mu_k \cos 35°)]}{m_1 + m_2}$

$$= \frac{9.81 \text{ m/s}^2 [3.0 \text{ kg} - (1.5 \text{ kg})(\sin 35° + 0.40 \times \cos 35°)]}{4.5 \text{ kg}} = 3.6 \text{ m/s}^2.$$

6-37.   The force exerted by the spring is $\mathbf{F} = -k\mathbf{x}$, where $\mathbf{x}$ is the displacement of the end of the spring. By Newton's Third Law the force required to change the length of the spring is equal in magnitude and opposite in direction to the force exerted by the spring. Therefore the magnitude of the force required to stretch the spring to twice its length is: $|F| = k|x| = 150 \text{ N/m} \times 0.15 \text{ m} = $ 23 N.

To compress it to one-half its length, $x = (0.15 \text{ m})/2 = 0.075 \text{ m}$. The magnitude of this force is $|F| = kx = 150 \text{ N/m} \times 0.075 \text{ m} =$ 11 N.

6-39.   In stretching from 6.3 to 1.2 cm, $k = \dfrac{1.0 \text{ N}}{(10.2 - 6.3) \text{ cm}} =$ 0.26 N/cm = 26 N/m. In stretching from 6.3 to 16.5 cm, $k = \dfrac{2.0 \text{ N}}{(16.5 - 6.3) \text{ cm}} =$ 0.20 N/cm = 20 N/m.   Since k is not constant, the spring does not obey Hooke's Law.

6-51.   The motion described in this problem isn't really uniform circular motion because the woman's speed varies during the swing. However, she is moving in a circular arc, so the centripetal force acting on her at any point is the swing is still given by $F = \dfrac{mv^2}{r}$, where v is her instantaneous speed . Strictly speaking, the forces acting on the woman are her weight and the normal force exerted by the seat. However, if we neglect the mass of the seat, then the normal force will be the same as the total tension in the two ropes that support the seat. So at the bottom of the swing, the net vertical force acting on the woman is $2T - mg$, which must be the centripetal force at that point: $2T - mg = \dfrac{mv^2}{r} \Rightarrow T = m\dfrac{v^2 + gr}{2r}$. The radius is the length of the rope (5.0 m), and her speed is 5.0 m/s, so $T = (60 \text{ kg})\dfrac{(5.0 \text{ m/s})^2 + (9.81 \text{ m/s}^2)(5.0 \text{ m})}{2(5.0 \text{ m})} =$ 4.4 × 10² N.

6-53. The net force in the vertical direction is zero because the car doesn't move in that direction. However, the net horizontal force cannot be zero because a centripetal force is needed to make the car go in a circle. That force comes from the horizontal component of the normal force. From the free-body diagram, $N \cos \theta = mg$, $N \sin \theta = mv^2/r$. (Note that even though the diagram looks very much like one for an object on an inclined plane, the motion being described is quite different and requires a different approach to the analysis.) Dividing the second equation by the first gives $\tan \theta = \dfrac{v^2}{rg}$. The required speed is 75 kph = 20.8 m/s. Then

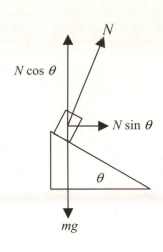

$$\theta = \tan^{-1}\left(\frac{v^2}{rg}\right) = \tan^{-1}\left(\frac{(20.8 \text{ m/s})^2}{(400 \text{ m})(9.81 \text{ m/s}^2)}\right) = \underline{6.3°}$$

6-55. Static friction provides the centripetal force that keeps the ant moving in a circle. The maximum force the surface can exert will occur at the maximum distance the ant reaches before sliding, which is $f_{s,max} = \mu_s N = \mu_s mg = \dfrac{mv^2}{r_{max}}$. The speed is the circumference of the circle divided by the time T for one revolution, and T is the reciprocal of the revolution rate $f \Rightarrow v = \dfrac{2\pi r}{T} = 2\pi rf$. Combining all these

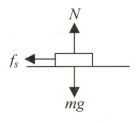

gives $r_{max} = \dfrac{\mu_s g}{4\pi^2 f^2} = \dfrac{(0.30)(9.81 \text{ m/s}^2)}{4\pi^2 (45 \text{ rev/min})^2 \left(\dfrac{1 \text{ min}}{60 \text{ s}}\right)^2} = \underline{0.13 \text{ m}}$

<u>(13 cm)</u>.

6-57. The time T for one complete orbit is 24 h = 8.64×10^4 s, and the satellite's speed is the circumference of its orbit divided by T. The "weight" is the centripetal force, so

$$w = \frac{mv^2}{r} = \frac{m\left(\dfrac{2\pi r}{T}\right)^2}{r} = \frac{4\pi^2 mr}{T^2} = \frac{4\pi^2 (1 \text{ kg})(4.23 \times 10^7 \text{ m})}{(8.64 \times 10^2 \text{ s})^2} = \underline{0.224 \text{ N}}.$$

6-59. The vertical component of the tension in the cable balances the weight of the rider, and the horizontal component provides the centripetal acceleration. Call the mass of the rider m and the cable tension T. Then

$T\sin\theta = \dfrac{mv^2}{r}$; $T\cos\theta = mg$, where v is the tangential

speed of the rider. Dividing the first equation by the second

gives $\tan\theta = \dfrac{v^2}{gr}$. The radius r is $(3 + 7\sin\theta)$ m, so

$\dfrac{v^2}{g(3+7\sin\theta)} = \tan\theta$. This gives $v = \sqrt{g\tan\theta(3+7\sin\theta)}$.

For $\theta = 35°$, $v = \underline{6.9 \text{ m/s}}$.

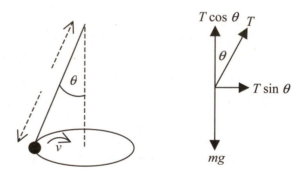

6-65. A free body diagram is shown. T is the tension in the string and v is the speed of the ball as it moves in the circle. The sum of forces in the vertical direction is zero, and the horizontal component of the tension provides the centripetal force that keeps the ball moving in

its circle. Thus $T\cos\theta = mg$, $T\sin\theta = \dfrac{mv^2}{r}$,

where $r = l\sin\theta$ is the radius of the circle. Divide the second equation by the first:

$\dfrac{\sin\theta}{\cos\theta} = \tan\theta = \dfrac{v^2}{gr}$. This means the speed of

the ball must satisfy $v = \sqrt{gr\tan\theta}$.

Substituting for r gives $v = \sqrt{gl\sin\theta\tan\theta}$.

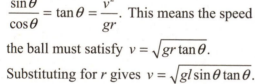

6-75. In Problem 6-60, an equation was derived relating the speed of the mass at the end of the pendulum and the angle the pendulum makes with the vertical. Using the variables in this

problem gives $\tan\alpha = \dfrac{v^2}{g(R+l\sin\alpha)}$. However, the speed we need is v_0, which is the speed at a

distance R from the center. To figure out v_0 in terms of v, we must recognize that the time for one complete revolution is the same for all parts of the apparatus. This time is the circumference of

one of the circles divided by the corresponding speed, so $\dfrac{2\pi R}{v_0} = \dfrac{2\pi(R+l\sin\alpha)}{v}$, from which we

get $v = \dfrac{R+l\sin\alpha}{R}v_0$. Then $\tan\alpha = \dfrac{(R+l\sin\alpha)^2 v_0^2}{gR^2(R+l\sin\alpha)} = \dfrac{(R+l\sin\alpha)v_0^2}{gR^2}$, which gives

$\underline{v_0^2 = \dfrac{gR^2\tan\alpha}{R+l\sin\alpha}}$. For the numerical data given, $v_0 = \sqrt{\dfrac{(9.81 \text{ m/s}^2)(0.20 \text{ m})^2(\tan 45°)}{0.20 \text{ m} + (0.30 \text{ m})(\sin 45°)}} = \underline{2.2 \text{ m/s}}$.

6-77. $f_{s,max} = \mu_s N = \mu_s mg.$

$\sum F_x = -\mu_s mg = ma_x \Rightarrow a_x = -\mu_s g.$

$v^2 - v_0^2 = 2a_x x \Rightarrow x = \dfrac{v^2 - v_0^2}{2a_x} = -\dfrac{v_0^2}{2a_x} = \dfrac{v_0^2}{2\mu_s g}.$ $v_0 = 90$

kph $= 25$ m/s.

$x = \dfrac{(25 \text{ m/s})^2}{2(0.80)(9.81 \text{ m/s}^2)} = \underline{40 \text{ m}}.$

$t = \dfrac{v - v_0}{a_x} = \dfrac{v_0}{\mu_s g} = \dfrac{25 \text{ m/s}}{(0.80)(9.81 \text{ m/s}^2)} = \underline{3.2 \text{ s}}.$

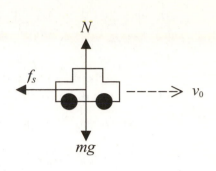

6-79.   (a) The free-body diagram is shown at the right. It is assumed that the initial velocity points to the right ($+x$), so the kinetic friction force points left ($-x$).

(b) $\sum F_y = N - mg = 0 \Rightarrow N = mg = (40 \text{ kg})(9.81 \text{ m/s}^2) = 392$ N.

Final result $\mathbf{N} = 3.9 \times 10^2 \mathbf{j}$ N.

(c) $f_k = \mu_k N = (0.80)(392 \text{ N}) = 314$ N. Final result

$\mathbf{f}_k = -3.1 \times 10^2 \mathbf{i}$ N.

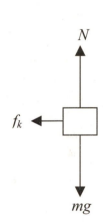

(d) $\mathbf{w} = -mg\mathbf{j} = \underline{3.9 \times 10^2 \ \mathbf{j}}$ N. $\mathbf{F}_{net} = \sum F_x \mathbf{i} + \sum F_y \mathbf{j} = -f_k \mathbf{i} = -314\mathbf{i}$ N.

Final result $\mathbf{F}_{net} = \underline{-3.1 \times 10^2 \mathbf{i}}$ N.

(e) $\mathbf{a} = \dfrac{\mathbf{F}_{net}}{m} = \dfrac{-314\mathbf{i} \text{ N}}{40 \text{ kg}} = -7.85\mathbf{i}$ m/s^2. Final result $\mathbf{a} = \underline{-7.9\mathbf{i} \text{ m/s}^2}$.

$x = \dfrac{v^2 - v_0^2}{2a_x}.$ $v_0 = 80$ kph $= 22.2$ m/s. $x = \dfrac{0 - (22.2 \text{ m/s})^2}{2(-7.85 \text{ m/s}^2)} = \underline{31 \text{ m}}.$

6-83.

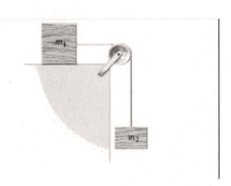

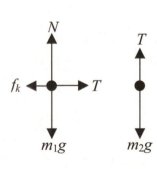

(a) Take the x direction to point right and the y direction to point up. Newton's Second Law for m_1 gives $\sum F_y = N - m_1 g = 0$, $\sum F_x = T - f_k = m_1 a_1$. For m_2, $\sum F_y = T - m_2 g = m_2 a_2$. We see that $N = m_1 g \Rightarrow f_k = \mu_k m_1 g \Rightarrow T - m_1 g = m_1 a_1$. If the string passing over the pulley does not stretch, then the magnitude of a_2 must be the same as the magnitude of a_1. If m_1 moves in the $+x$ direction, then m_2 must move in the $-y$ direction, which means $a_2 = -a_1$. Then the equations for $\sum F_x$ for m_1 and $\sum F_y$ for m_2 can be combined to give an equation for a_1: $a_1 = \dfrac{m_2 - \mu_k m_1}{m_1 + m_2} g$.

6-85.   The wording in the problem implies the springs are in series with one spring hanging from the end of the other. Suppose spring 1 has force constant $k_1 = 2.0 \times 10^3$ N/m and spring 2 has force constant $k_2 = 3.0 \times 10^3$ N/m. Let us further suppose that spring 1 is at the bottom, with the mass m hanging from it. We assume that the mass and the springs are in equilibrium, so the force stretching spring 1 is mg. Then spring 1 stretches Δx_1 because of that force. Spring 1 exerts a force on spring 2 causing it to stretch by Δx_2. If the springs and the mass are in equilibrium, then the net force on each object must be zero. Since there is a force mg pulling down on spring 1, this same force must be pulling on spring 2. Then for each spring, $mg = k_1 \Delta x_1$ and $mg = k_2 \Delta x_2$, respectively. The total stretch of both springs is $\Delta x = \Delta x_1 + \Delta x_2 = \dfrac{mg}{k_1} + \dfrac{mg}{k_2} = mg\left(\dfrac{1}{k_1} + \dfrac{1}{k_2}\right)$.

This means the combined springs act like a single spring with an effective force constant given by $\Delta x = \dfrac{mg}{k} = mg\left(\dfrac{1}{k_1} + \dfrac{1}{k_2}\right)$, or $\dfrac{1}{k} = \dfrac{1}{k_1} + \dfrac{1}{k_2}$.

For the springs in this problem, $k = \dfrac{k_1 k_2}{k_1 + k_2} = \dfrac{6 \times 10^6 \text{ N}}{5 \times 10^3 \text{ m}} = 1.2 \times 10^3 \dfrac{\text{N}}{\text{m}}$. The magnitude of the total force stretching the "spring" is $mg = (5 \text{ kg})(9.81 \text{ m/s}^2) = 49.1$ N. The total stretch of both springs is $\Delta x = \dfrac{|F|}{k} = \dfrac{49.1 \text{ N}}{1.2 \times 10^3 \text{N/m}} = 0.0409$ m (4.1 cm). The individual stretches are

$\Delta x_1 = \dfrac{|F|}{k_1} = \dfrac{49.1 \text{ N}}{2.0 \times 10^3 \text{N/m}} = \underline{2.5 \text{ cm}}$, and $\Delta x_2 = \dfrac{|F|}{k_2} = \dfrac{49.1 \text{ N}}{3.0 \times 10^3 \text{N/m}} = \underline{1.6 \text{ cm}}$, which do indeed give a total stretch of 4.1 cm to two significant figures.

6-87.   (a) If the spring is massless, then the force the spring exerts on the block will be the same as the force exerted on the spring. To start the block moving, the pull P must exceed the total force pointing down the slope, which is the sum of the maximum static friction force $\mu_s N$ and the component of the block's weight parallel to the slope. Adding forces normal to the slope gives $\sum F_y = N - mg\cos 30° = 0 \Rightarrow N = mg\cos 30°$

$= (1.5 \text{ kg})(9.81 \text{ m/s}^2)\cos 30°$

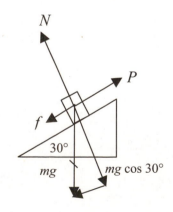

$= (14.7 \text{ N})\cos 30° = 12.7 \text{ N}$. Parallel to the slope, $\sum F_x = P - mg\sin 30° - \mu_s N = 0$

$\Rightarrow P = (14.7 \text{ N})\sin 30° + (0.60)(12.7 \text{ N}) = 15.0 \text{ N}$. The stretch of the spring under this force will

be $\Delta x = \dfrac{|P|}{k_1} = \dfrac{15.0 \text{ N}}{1.2 \times 10^3 \text{N/m}} = 1.2 \times 10^{-2} \text{ m (1.2 cm)}$.

(b) When the block begins to move, the force changes from static friction to kinetic friction. Now

$P - mg\sin 30° - \mu_k N = ma_x$, which gives $a_x = \dfrac{P - mg\sin 30° - \mu_k N}{m}$.

$= \dfrac{15.0 \text{ N} - (14.7 \text{ N})\sin 30° - (0.40)(12.7 \text{ N})}{1.5 \text{ kg}} = 1.7 \text{ m/s}^2$. The positive result means the

acceleration points up the incline.

(c) "Constant speed" means $a_x = 0$, so $P = mg\sin 30° + \mu_k N = 7.34 \text{ N} + (0.40)(12.7 \text{ N}) = 12.4 \text{ N}$.

Now the stretch of the spring is $\Delta x = \dfrac{|P|}{k_1} = \dfrac{12.4\text{N}}{1.2 \times 10^3 \text{N/m}} = \underline{1.0 \times 10^{-2} \text{ m (1.0 cm)}}$.

CHAPTER 7 WORK AND ENERGY

7-1. The work done by a force F is $W = Fs \cos \theta$, where s is the displacement and the angle θ is the angle between the force and the displacement vectors. Since the force applied to the car is in the same direction as the displacement, the angle is 0° and $\cos 0° = 1$.

$W = Fs \cos \theta = (300 \text{ N})(5.0 \text{ m})\cos 0° = \underline{1500 \text{ J}} = \underline{1.5 \times 10^3 \text{ J}}$

7-3. With each movement, either forward or backward, the displacement and the force are in the same direction. Therefore, positive work is done on the saw.

$W = F_x \Delta x = (35 \text{ N})[(30 \times 0.12 \text{ m}) + (30 \times 0.12 \text{ m})] = \underline{252 \text{ J}}$

7-7. $W = F \Delta y = mg \Delta y = (75 \text{ kg})(9.81 \text{ m/s}^2)(320 \text{ m}) = \underline{235\ 000 \text{ J}}$

The average rate of doing work is

$\dfrac{W}{\Delta t} = \dfrac{235\ 000 \text{ J}}{659 \text{ s}} = \underline{357 \text{ J/s}}$.

7-9. The first tugboat does positive work W_1 on the barge, where

$W_1 = F_1 s \cos \theta_1 = (2.5 \times 10^5 \text{ N})(100 \text{ m}) \cos 30° = \underline{2.2 \times 10^7 \text{ J}}$.

Similarly, the second boat does positive work W_2 on the barge, where

$W_2 = F_2 s \cos \theta_2 = (1.0 \times 10^5 \text{ N})(100 \text{ m}) \cos 15° = \underline{1.0 \times 10^7 \text{ J}}$.

The total work done by both tugboats is then $W = W_1 + W_2 = \underline{3.2 \times 10^7 \text{ J}}$.

7-11. P is the push applied by the man. From the free-body diagram, $N = mg \cos \theta$ and $P = mg \cos \theta + f$, where $f = \mu_k N = \mu_k mg \cos \theta$ is the frictional force. Therefore, $P = mg \cos \theta + \mu_k mg \cos \theta$. To raise the box to a height h, the box is pushed up the incline a distance $s = h/\sin \theta$. The work done on the box by the man is $W = Ps = (mg \cos \theta + \mu_k mg \cos \theta)(h/\sin \theta)$, which may be simplified to

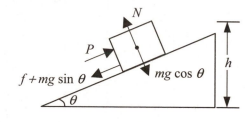

$W = mgh(1 + \mu_k \cot \theta) = (60 \text{ kg})(9.81 \text{ m/s}^2)(2.5 \text{ m})[1 + (0.45) \cot 30°] = \underline{2.6 \times 10^3 \text{ J}}$.

7-17. (a) If the cart is to move at constant velocity, the magnitude of the applied force T_1 must equal that of the kinetic frictional force f_k, which is oppositely directed. The work the man does on the cart is $W_1 = T_1 \Delta x = (250 \text{ N})(50 \text{ m}) = \underline{1.3 \times 10^4 \text{ J}}$.

(b) When the force is applied in the direction shown, work is done by the horizontal component of the tension $\mathbf{T}$. Because the vertical component of the tension would reduce the normal force and also the frictional force.

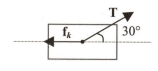

In this problem, additional mass has been added so that the frictional force is the same in both cases. To determine the tension, we assume that the resulting horizontal acceleration is equal to zero, so that $T_2 \cos 30° - f_k = 0$. Solving for T gives,

$T_2 = f_k/(\cos 30°) = (250 \text{ N})/(\cos 30°) = \underline{290 \text{ N}}$.

The work the man must now do is

$W_2 = T(\cos 30°)\Delta x = (290 \text{ N})(\cos 30°)(50 \text{ m}) = \underline{1.3 \times 10^4 \text{ J}}$, which is the same as W_1.

7-19.   (a)   The net force on the elevator is $T_1 - Mg = Ma$, where T_1 is the tension in the elevator cable and M is the elevator mass including its load.
$T_1 = M(g + a) = (1200 \text{ kg})(9.81 \text{ m/s}^2 + 1.5 \text{ m/s}^2) = \underline{1.4 \times 10^4 \text{ N}}$
For the counterweight, the tension is
$T_2 = m(g - a) = (1000 \text{ kg})(9.81 \text{ m/s}^2 - 1.5 \text{ m/s}^2) = \underline{8.3 \times 10^3 \text{ N}}$.
(b)   To determine the distance traveled in the 1.0 s interval, apply the kinematic relation,
$y = \frac{1}{2}at^2 = \frac{1}{2}(1.5 \text{ m/s}^2)(1.0 \text{ s}) = 0.75 \text{ m}$.

Assuming the elevator is moving upward during the interval, the work done by the pulley motor is

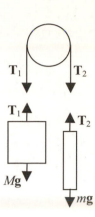

$W = T_1 y + T_2 y = (1.4 \times 10^4 \text{ N})(0.75 \text{ m}) - (8.3 \times 10^3 \text{ N})(0.75 \text{ m}) = \underline{4300 \text{ J}}$.
(c)   For motion at constant speed, $T_1 = Mg = (1200 \text{ kg})(9.81 \text{ m/s}^2) = \underline{1.2 \times 10^4 \text{ N}}$ and
$T_2 = mg = (1000 \text{ kg})(9.81 \text{ m/s}^2) = \underline{9.8 \times 10^3 \text{ N}}$
The work done during the upward displacement is
$W = (T_1 - T_2)y = (1.2 \times 10^4 - 9.8 \times 10^3)(10.0 \text{ m}) = \underline{2.2 \times 10^4 \text{ J}}$.

7-21.   (a)   The angle the rope makes with respect to the horizontal direction is $\theta = \sin^{-1}(10 \text{ m}/50 \text{ m})$. Since the boat is stationary, the horizontal component of the tension must be equal in magnitude to the wind force. Thus, $T \cos\theta = 7000 \text{ N}$, which can be solved for the tension
$$T = \frac{7000 \text{ N}}{\cos\theta} = \frac{7000 \text{ N}}{\cos\left[\sin^{-1}\left(\dfrac{10 \text{ m}}{50 \text{ m}}\right)\right]} = \underline{7100 \text{ N}}.$$

(b)   Since the wind force is 7000 N, the sailors must exert 7000 N in the direction of motion assuming the boat is to be pulled at constant speed. If 30 m of rope is pulled in, the hypotenuse is reduced to 20 m. The horizontal distance is found from the Pythagorean theorem,
$x_2 = \sqrt{(20 \text{ m})^2 - (10 \text{ m})^2} = 17 \text{ m}$
Similarly, the original distance was
$x_1 = \sqrt{(50 \text{ m})^2 - (10 \text{ m})^2} = 49 \text{ m}$
So, the net displacement of the boat is $(49 \text{ m} - 17 \text{ m}) = 32 \text{ m}$ and the work done is
$W = (7000 \text{ N})(32 \text{ m}) = \underline{2.2 \times 10^5 \text{ J}}$.
In this new position, the tension is
$$T = \frac{7000 \text{ N}}{\cos\theta} = \frac{7000 \text{ N}}{\cos\left[\sin^{-1}\left(\dfrac{10 \text{ m}}{20 \text{ m}}\right)\right]} = \underline{8100 \text{ N}}.$$

7-23.   The work is equal to the area under the F_x versus x curve. The work done in moving between 0 and 6 m is + 8 J. Between 6 and 8 m, the work done is –2 J. Therefore, the total work done is
$W = +8 \text{ J} - 2 \text{ J} = \underline{+6 \text{ J}}$.

7-25.   The spring force is a restoring force, which is directed so as to return the spring to its equilibrium position. In this problem, the mass attached to the end of the spring is displaced in the positive x direction, but as the mass passes $x = 0$ m, the spring force changes direction. Since the force is varying, the work done by the spring is found through integration.
$$W_s = \int_a^b F_x(x)dx = \int_{-0.20 \text{ m}}^{0.40 \text{ m}} -kx \ dx = -\frac{1}{2}kx^2\Big|_{-0.20 \text{ m}}^{0.40 \text{ m}}$$
$$= -\frac{1}{2}(440 \text{ N/m})[(0.40 \text{ m})^2 - (-0.20 \text{ m})^2] = \underline{-26 \text{ J}}$$

7-27. To stretch the spring from 0 to d, requires work $W_0 = -\dfrac{1}{2}kd^2$. The amount of work to further stretch the spring to $2d$ is

$$W = \int_d^{2d} F_x(x)dx = \int_d^{2d} -kx\, dx = -\tfrac{1}{2}kx^2 \Big|_d^{2d}$$
$$= -\tfrac{1}{2}k[(2d)^2 - (d)^2] = -\tfrac{3}{2}kd^2 = 3W_0.$$

If we now wish to generalize this for greater distances, the work done would be

$$W = \int_{Nd}^{(N+1)d} kx\, dx = \tfrac{1}{2}kx^2 \Big|_{Nd}^{(N+1)d}$$
$$= -\tfrac{1}{2}k[(N+1)^2 d^2 - (Nd)^2] = -\tfrac{1}{2}k(2N+1)d^2 = (2N+1)W_0.$$

7-29. In the vertical direction, the magnitude of the external force is equal to twice the vertical component of the restoring force of the spring F_y. Consider one-half of the spring as represented in the triangle below the figure to the right. The extension of one-half of the spring from its equilibrium length of $l/2$ is equal to $\Delta l = \sqrt{(l/2)^2 + y^2} - l/2$.

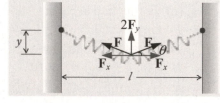

The applied force is

$$P = 2F\sin\theta = 2k(\Delta l)\frac{y}{\sqrt{(l/2)^2 + y^2}}.$$

where F is the magnitude of the force exerted by the spring. Substituting the equation for Δl gives

$$P = 2k\left[\sqrt{(l/2)^2 + y^2} - l/2\right]\frac{y}{\sqrt{(l/2)^2 + y^2}} = 2k\left[y - \frac{y(l/2)}{\sqrt{(l/2)^2 + y^2}}\right]$$

$$P = 2ky\left[1 - \frac{(l/2)}{\sqrt{(l/2)^2 + y^2}}\right].$$

The work done on the spring by the force P is

$$W = \int_0^y P\,dy = 2k\int_0^y y\left[1 - \frac{(l/2)}{\sqrt{(l/2)^2 + y^2}}\right]dy$$

$$= 2k\left\{\frac{y^2}{2} - \frac{l}{2}\left[\sqrt{(l/2)^2 + y^2} - (l/2)\right]\right\} = k\left[y^2 + \frac{l^2}{2} + \frac{l}{2}\sqrt{(l/2)^2 + y^2}\right].$$

As a check, the work done by the external force on the spring in displacing it by a distance y and extending it by Δl is equal to the change in the spring potential energy, $W = \tfrac{1}{2}k(\Delta l)^2$.

Then, the total work done on the whole spring is twice this value,

$$W = 2\left(\frac{1}{2}k(\Delta l)^2\right) = k\left[\sqrt{(l/2)^2 + y^2} - l/2\right]^2$$

which is the same result obtained when the quantity in the square brackets is expanded.

7-31. Consider the path from the origin to the point (2 m, 2 m, 0). This is a straight line path on which $x = y$ and, therefore, $dx = dy$. The work done on this path is

$$W = \int_{x_1}^{x_2} F_x \, dx + \int_{y_1}^{y_2} F_y \, dy + \int_{z_1}^{z_2} F_z \, dz$$

$$= \int_0^{x_2} (4x^2 + 1) \, dx + \int_0^{y_2} 2x \, dy = \int_0^{2\,m} (4x^2 + 1) \, dx + \int_0^{2\,m} 2x \, dx$$

$$= \tfrac{4}{3} x^3 + x + x^2 \Big|_0^{2\,m} = \underline{17 \text{ J}}.$$

7-33. (a) When the two atoms are close to each other, the force is repulsive. At longer distances, the force is attractive; and at the equilibrium separation, the force is zero. Setting the force equal to zero and solving for x, gives

$$F = Ax^{-13} - Bx^{-7} = 0$$

$$x = \sqrt[6]{\frac{A}{B}}.$$

(b) The work done is equal to the

$$W = \int_{x_{eq}}^{\infty} F_x \, dx = \int_{x_{eq}}^{\infty} Ax^{-13} - Bx^{-7} \, dx = A\left(\frac{x^{-12}}{12}\right) - B\left(\frac{x^{-6}}{6}\right)\Big|_{x_{eq}}^{\infty} = \frac{B^2}{12A} - \frac{B^2}{6A} = \underline{-\frac{B^2}{12A}}.$$

7-35. The Earth-Sun distance is $R = 1.5 \times 10^{11}$ m and the orbit is approximately circular, so the circumference of the Earth's orbit is $C = 2\pi R = 9.4 \times 10^{11}$ m. The Earth travels this distance in one year $= 3.156 \times 10^7$ s, so the average speed of the Earth is $v = C/t = 3.0 \times 10^4$ m/s. The Earth's mass is 5.98×10^{24} kg. The kinetic energy of the Earth in its orbit is

$$K = \tfrac{1}{2} mv^2 = \tfrac{1}{2}(5.98 \times 10^{24} \text{ kg})(3.0 \times 10^4 \text{ m/s})^2 = \underline{2.7 \times 10^{33} \text{ J}}.$$

7-39. (a) 80 km/hr $= 22.2$ m/s, so $K = \tfrac{1}{2} mv^2 = \dfrac{1}{2}(1600 \text{ kg})(22.2 \text{ m/s})^2 = \underline{4.0 \times 10^5 \text{ J}}$

(b) 20 km/hr $= 5.56$ m/s, so $K = \tfrac{1}{2} mv^2 = \dfrac{1}{2}(1600 \text{ kg})(5.56 \text{ m/s})^2 = \underline{2.5 \times 10^4 \text{ J}}$

(c) 140 km/hr $= 38.9$ m/s, so $K = \tfrac{1}{2} mv^2 = \dfrac{1}{2}(1600 \text{ kg})(38.9 \text{ m/s})^2 = \underline{1.2 \times 10^6 \text{ J}}$

7-43. $K_{Ball} = \tfrac{1}{2} m_{Ball} v_{Ball}^2 = \tfrac{1}{2}(0.045 \text{ kg})(45 \text{ m/s})^2 = \underline{46 \text{ J}}$

$K_{Person} = \tfrac{1}{2} m_{Person} v_{Person}^2 = \tfrac{1}{2}(75 \text{ kg})(1.0 \text{ m/s})^2 = \underline{38 \text{ J}}$

The kinetic energies of the golf ball and person are on the same order of magnitude.

7-45. The kinetic energy of the object as it is launched is $K = \tfrac{1}{2} mv^2 = \tfrac{1}{2}(0.150 \text{ kg})(5.0 \text{ m/s})^2 = \underline{1.9 \text{ J}}$.

The work done on the object is equal to the change in kinetic energy of the object. Since the object started from rest, the work is

$W = \Delta K = \underline{1.9 \text{ J}}$.

This energy was also the stored energy in the compressed spring. Therefore,

$$x = \sqrt{\frac{2W}{k}} = \sqrt{\frac{2(1.9 \text{ J})}{20 \text{ N/m}}} = \underline{0.44 \text{ m}}.$$

7-49.   Let d be the depth the bullet penetrates when it is fired with a speed v. Since d is assumed to be proportional to the kinetic energy, we may write the relationship as $d = kv^2$, where k is some constant. The two cases may then be compared to find the speed of the second bullet.

$d_1 = kv_1^2$ or 0.8 cm $= k(160$ m/s$)$

$d_2 = kv_2^2$ or 1.2 cm $= k v_2^2$

Taking the ratio of these two equations yields, $v_2 = 160\sqrt{\dfrac{1.2}{0.9}} = \underline{196\ \text{m/s}}$.

7-51.   The stored potential energy in the spring is equal to the work done in moving it to its initial position, $W = \frac{1}{2}kx_1^2$. After the cantilever is released, the cantilever has both kinetic energy and spring potential energy, the sum of which equals the work done. Therefore,

$\frac{1}{2}kx_1^2 = K + \frac{1}{2}kx_2^2$

$K = \frac{1}{2}k(x_1^2 - x_2^2) = \frac{1}{2}(2.5 \times 10^{-2}\ \text{N/m})[(3.0 \times 10^{-8}\ \text{m})^2 - (2.5 \times 10^{-8}\ \text{m})^2]$

$= \underline{3.4 \times 10^{-18}\ \text{J}}$

7-53.   (a) $\Delta K = K_2 - K_1 = \frac{1}{2}(40\ \text{kg})(2.0\ \text{m/s})^2 = \underline{80\ \text{J}}$

(b) $W_f = \mathbf{f}_k \cdot \mathbf{x} = -f_k x = -\mu_k N x$

Applying Newton's second law in the vertical direction, $N = mg$. The sum of the horizontal forces is $P - f_k = ma$ or

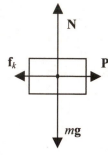

$a = \dfrac{P - \mu_k mg}{m} = \dfrac{250\ \text{N} - (0.60)(40\ \text{kg})(9.81\ \text{m/s}^2)}{(40\ \text{kg})} = 0.36\ \text{m/s}^2.$

We can find the sliding distance from $x = \dfrac{v^2 - v_0^2}{2a} = \dfrac{(2\ \text{m/s})^2 - 0}{2(0.36\ \text{m/s}^2)} = 5.5\ \text{m}.$

The work done by the friction force is

$W_f = -\mu_k N x = -\mu_k mgx = -(0.60)(40\ \text{kg})(9.81\ \text{m/s}^2)(5.5\ \text{m}) = \underline{-1295\ \text{J}}.$

(c) $W_w = \mathbf{P} \cdot \mathbf{x} = Px = (250\ \text{N})(5.5\ \text{m}) = \underline{1375\ \text{J}}.$

Note that the total work done on box is $W = W_w + W_f = 80\ \text{J} = \Delta K.$

7-55. $W = mg\Delta y = (75\ \text{kg})(9.81\ \text{m/s}^2)(10\ \text{m}) = \underline{7.4 \times 10^3\ \text{J}}$

The metabolization of one apple is $4.6 \times 10^5\ \text{J}$, so in climbing the stairs $(7.4 \times 10^3\ \text{J})/(4.6 \times 10^5\ \text{J}) = \underline{1.6\ \%}$ of the energy stored in an apple is used.

7-57.   The gravitational potential energy of the water is $U = mgh$. The mass of the water can be determined from its density $\rho = m/V = 1000\ \text{kg/m}^3$. To determine the volume, combine these two relations,

$V = \dfrac{m}{\rho} = \dfrac{U}{\rho gh} = \dfrac{2.0 \times 10^{13}\ \text{J}}{(1000\ \text{kg/m}^3)(9.81\ \text{m/s}^2)(250\ \text{m})} = \underline{8.2 \times 10^6\ \text{m}^3}$

7-59.   Applying the principle of the conservation of mechanical energy to this situation

$\frac{1}{2}mv_1^2 + mgh_1 = \frac{1}{2}mv_2^2 + mgh_2$

$\frac{1}{2}mv_1^2 = mgh_2$

$h_2 = \dfrac{v_1^2}{2g} = \dfrac{(10\ \text{m/s})^2}{2(9.81\ \text{m/s}^2)} = \underline{5.1\ \text{m}}$

However, the athlete actually reaches a height of 5.7 m. In the process of performing the vault, the athlete exerts an additional amount of internal energy in pulling himself upward that results in the additional 0.6 m height.

7-61.   Use conservation of energy, since friction effects are negligible.

$\frac{1}{2}mv_1^2 + mgh_1 = \frac{1}{2}mv_2^2 + mgh_2$

$mgh_1 = \frac{1}{2}mv_2^2$ which gives $v_2 = \sqrt{2gh_1} = \sqrt{2(9.81 \text{ m/s}^2)(500 \text{ m})} = \underline{99 \text{ m/s}}$

The kinetic energy of the 2.0×10^7 kg of snow is

$K = \frac{1}{2}mv^2 = \frac{1}{2}(2.0 \times 10^7 \text{ kg})(99 \text{ m/s})^2 = \underline{9.8 \times 10^{10} \text{ J}}$.

The equivalent amount of TNT is

$N = \dfrac{9.8 \times 10^{10} \text{ J}}{4.2 \times 10^9 \text{ J/ton}} = \underline{23 \text{ tons of TNT}}$.

7-63.   Consider the free-body diagram superposed on the drawing.
In the direction perpendicular to the inclined plane, $N - mg$
$\cos\theta = 0$ or $N = mg \cos\theta$. The distance the block slides is
$s = (1.5 \text{ m})/\sin(15°) = 5.8 \text{ m}$; and the work done by the
frictional force on the block is

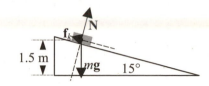

$W = E_2 - E_1 = (K_2 + U_2) - (K_1 + U_1)$
$\quad = K_2 - U_1 = -fs$

which is then used to find the coefficient of kinetic friction, since $f_k = \mu_k N$.

$mgh_1 - \frac{1}{2}mv_2^2 = \mu_k Ns$

$\mu_k = \dfrac{mgh_1 - \frac{1}{2}mv_2^2}{Ns} = \dfrac{mgh_1 - \frac{1}{2}mv_2^2}{(mg\cos\theta)s} = \dfrac{2gh_1 - v_2^2}{2(g\cos\theta)s} = \dfrac{2(9.81 \text{ m/s}^2)(1.5 \text{ m}) - (3.5 \text{ m/s})^2}{2(9.81 \text{ m/s}^2)(\cos15°)(5.8 \text{ m})} = \underline{0.16}$

7-65.   The difference in the total energy between the release from rest of the ball and its landing is equal
to the amount of work done by frictional forces, $W = \Delta E = (K_2 - U_2) - (K_1 - U_1)$. The fraction lost
to air friction is equal to W/U_1.

$W = K_2 - U_1 = \frac{1}{2}mv_2^2 - mgh_1$

$\dfrac{W}{U_1} = \left| \dfrac{\frac{1}{2}mv_2^2 - mgh_1}{mgh_1} \right| = \left| \dfrac{v_2^2 - 2gh_1}{2gh_1} \right| = \left| \dfrac{(9.0 \text{ m/s})^2 - 2(9.81 \text{ m/s}^2)(20 \text{ m})}{2(9.81 \text{ m/s}^2)(20 \text{ m})} \right| = \underline{0.79 \text{ or } 79\%}$.

7-67.   Since frictional effects are ignored, conservation of energy applies.

$\frac{1}{2}mv_1^2 + mgh_1 = \frac{1}{2}mv_2^2 + mgh_2$

$mgh_1 = \frac{1}{2}mv_2^2 + mgh_2$

$v_2 = \sqrt{2g(h_1 - h_2)} = \sqrt{2(9.81 \text{ m/s}^2)(3.0 \text{ m})} = \underline{7.7 \text{ m/s}}$

7-69. When the ball is displaced from its equilibrium position, it is raised to a
height h relative to its lowest position before it is released from rest. To
find the height h, a horizontal line is drawn from the ball's initial position
to the vertical line. A right triangle is then formed, from which we can
find the length of the side labeled b. The initial height is then,
$h = (10 \text{ m}) - b = (10 \text{ m}) - (10 \text{ m}) \cos 35° = (10 \text{ m})(1 - \cos 35°)$.
We can then apply the principle of conservation of mechanical energy to
find the ball's kinetic energy at its lowest point where its potential energy
$U_2 = 0$.
$K_2 + U_2 = K_1 + U_1$
$K_2 = U_1$
$K_2 = mgh_1 = (600 \text{ kg})(9.81 \text{ m/s}^2)(10 \text{ m})(1 - \cos 35°) = \underline{1.1 \times 10^4 \text{ J}}$

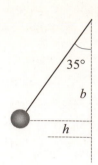

7-71. At position **1**, choosing the downward direction as positive, the
sum of the forces on the stone is

$T_1 + mg = \dfrac{mv_1^2}{r}$

Since the tension is about zero, the speed at the top of the circle is
$v_1^2 = rg$. Therefore, the total energy is

$E = K_1 + U_1 = \frac{1}{2}mv_1^2 + mgh_1 = \frac{1}{2}mrg + 2mrg = \frac{5}{2}mrg$

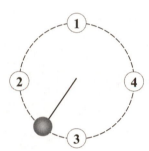

At position **3**, the tension is directed upward and the sum of the forces on the stone is

$T_3 - mg = \dfrac{mv_3^2}{r}$. To find v_3, consider the total energy at position **3**.

$E = K_3 + U_3 = \frac{1}{2}mv_3^2 + mgh_3 = \frac{1}{2}mv_3^2 = \frac{5}{2}mrg$

$v_3^2 = 5rg$

Therefore, the tension at position **3** is

$T_3 = mg + \dfrac{mv_3^2}{r} = mg + \dfrac{m5rg}{r} = 6mg = 6(0.90 \text{ kg})(9.81 \text{ m/s}^2) = \underline{53 \text{ N}}$.

7-79. The particle will slide along the sphere until the normal force N
goes to zero. At the angular position shown, the sum of the forces
on the particle in the radial direction is $mg \cos \theta - N = mv^2/R$,
which gives $v^2 = gR \cos \theta$. The conservation of energy is then
applied to the situation,
$mgR = mgR \cos \theta + \frac{1}{2}mv^2 = mgR \cos \theta + \frac{1}{2}m(gR \cos \theta)$..
Solving for θ gives, $\theta = \cos^{-1}\left(\frac{2}{3}\right) = \underline{48.2°}$.

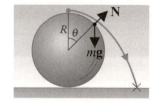

7-85.   (a) Frictional forces do work on the car over the stopping distance, $W = -fs$. This results in a change in kinetic energy ΔK.

$$W = -\mu mgs = \Delta K = \tfrac{1}{2}mv_2^2 - \tfrac{1}{2}mv_1^2$$

$$s = \frac{0 - \tfrac{1}{2}mv_1^2}{-\mu mg} = \frac{\tfrac{1}{2}(25 \text{ m/s})^2}{(0.90)(9.81 \text{ m/s}^2)} = \underline{35.4 \text{ m}}$$

To determine the acceleration, assuming the acceleration is constant as the car is slowing, we can use the kinematic relation

$$v_2^2 - v_1^2 = 2as.$$

$$a = \frac{v_2^2 - v_1^2}{2s} = \frac{0 - (25 \text{ m/s})^2}{2(35.4 \text{ m})} = \underline{-8.8 \text{ m/s}^2}$$

(b) $f = \mu N = \mu mg = (0.90)(1200 \text{ kg})(9.81 \text{ m/s}^2) = \underline{1.06 \times 10^4 \text{ N}}$

$W = -fs = -(10\,600 \text{ N})(35.4 \text{ m}) = \underline{3.75 \times 10^5 \text{ J}}$

7-91.   The energy lost to friction is the difference between the final and initial mechanical energy of the luger. The initial mechanical energy is equal to the potential energy because the luger starts from rest.

$$\Delta E = mgh_1 - \left(\tfrac{1}{2}mv_2^2 + mgh_2\right)$$

$$= mg(h_1 - h_2) - \tfrac{1}{2}mv_2^2$$

$$= (95 \text{ kg})(9.81 \text{ m/s}^2)(350 \text{ m} - 240 \text{ m}) - \tfrac{1}{2}(95 \text{ kg})\left[(130 \text{ km/h})\left(\frac{1 \text{ h}}{3600 \text{ s}}\right)\left(\frac{1000 \text{ m}}{1 \text{ km}}\right)\right]^2$$

$$= \underline{4.1 \times 10^4 \text{ J}}$$

8-5. $U(x) = -\int_{x_0}^{x} F(x')dx' = -\int_{x_0}^{x} A(x')^3\, dx' = \dfrac{Ax^4}{4} - \dfrac{Ax_0^4}{4}$. Choose $x_0 = 0$. Then $U(x) = \dfrac{Ax^4}{4}$. The law

of conservation of energy states that if no nonconservative forces act, $U_2 + K_2 = U_1 + K_1$. Let

point 1 be x with $K_1 = 0$ (the particle is at rest) and point 2 be $x = 0$ with a kinetic energy

$K_2 = \dfrac{mv^2}{2}$. Then $0 + \dfrac{mv^2}{2} = \dfrac{Ax^4}{4} + 0 \Rightarrow v = \sqrt{\dfrac{Ax^4}{2m}} = \sqrt{\dfrac{(50\ \text{N/m}^3)(0.50\ \text{m})^4}{2(0.050\ \text{kg})}} = 5.6\ \text{m/s}.$

8-7. $U(x) = -\int_{0}^{x} F(x')dx' = -\int_{0}^{x}[-2x'-(x')^3]dx' = x^2 + \dfrac{x^4}{4}$, where it has been assumed that

$x_0 = 0$. (A note about units: the coefficient 2 of the x term in U has units of N/m, while the
coefficient of the x^3 term is actually 1 N/m^3. In the integral, dx has units of m, so the coefficient of
the x^2 term is actually 1 N/m, and the coefficient of the x^4 term is now 0.25 N/m^3. It's important to
remember that apparently abstract operations like integration can involve units.) The table that
follows evaluates the equation for the values of x requested in the problem.

x (m)	U (J)
1.0	1.3
2.0	8.0
3.0	29.3

8-13. Since **F** is a conservative force, the work done by it as a particle moves
from one point to another is independent of the path followed. The
path suggested by the hint makes evaluation of the integral easy. Part I
of the path goes from $(0,0)$ to $(x,0)$ (x varies while y is held constant at
0), and part II goes from $(x,0)$ to (x,y) (y varies while x is held
constant). Do the integral for part I of the path:

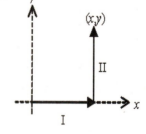

$$dW = \mathbf{F}\cdot d\mathbf{r} \Rightarrow W_{\mathrm{I}} = \int_{(0,0)}^{(x,0)}(0\mathbf{i} + bx'\mathbf{j})\cdot(dx'\mathbf{i} + dy'\mathbf{j}) = bx'\int_{y'=0}^{y'=0} dy' = 0$$

(since y doesn't change along part I of the path). Another interpretation
is that along I **F** points in the y direction and the displacement points in
the x direction. Since **F** is perpendicular to d**r** along I, the work must
be zero. Along part II of the path, we get

$$W_{\mathrm{II}} = \int_{(x,0)}^{(x,y)}(by'\mathbf{i} + bx\mathbf{j})\cdot(dx'\mathbf{i} + dy'\mathbf{j}) = bx\int_{y'=0}^{y'=y} dy' = bxy. \text{ The total work}$$

is $W = W_{\mathrm{I}} + W_{\mathrm{II}} = bxy$. The potential energy is $U = -W = -bxy$.

8-25. (a) K is a maximum when U is a minimum. From the graph in Example 4, this occurs at
$x \approx -5$ m. To find the exact location, differentiate $U(x)$ and set the derivative equal to zero:
$\dfrac{dU}{dx} = 687 + 150x = 0$, which gives $x = -\dfrac{687}{150}$ m $= -4.58$ m. This is in very good agreement

with the 1-digit estimate from the graph.

To find the speed at this point, use $E = U + K \Rightarrow K = E - U$. Substitute $x = -4.58$ m in the equation for U to get $U = -1573$ J. Then $K = 6180$ J $+ 1573$ J $= 7753$ J. Then $v = \sqrt{\dfrac{2K}{m}}$

$$= \sqrt{\frac{2(7753 \text{ J})}{70 \text{ kg})}} = \underline{14.9 \text{ m/s.}}$$

(b) The force is $F_x = -\dfrac{dU}{dx} = -687 - 150x$. For x within the range of the problem, the maximum force occurs at the turning point, which is $x = -14.7$ m. The magnitude of the force at this value of x is $F = 1.52 \times 10^3$ N, so the magnitude of the acceleration is

$$a = \frac{F}{m} = \frac{1.52 \times 10^3 \text{ N}}{70 \text{ kg}} = 21.7 \text{ m/s}^2, \text{ which is about 2.2 gees.}$$

8-29. $E = U + K = A|x| + K$. At $x = 0$, $K = \dfrac{mv^2}{2} \Rightarrow E = \dfrac{mv^2}{2}$. The

energy is constant. At the turning points, $K = 0$, which gives

$\dfrac{mv^2}{2} = A|x| + 0 \Rightarrow |x| = \dfrac{mv^2}{2A}$. The final result is that the turning

points are at $\underline{x = \pm \dfrac{mv^2}{2A}.}$

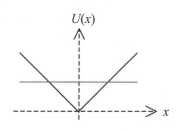

8-31. $U = -\dfrac{2.0}{1 + x^2}$. The graph was plotted using a spreadsheet.

At the turning points, $K = 0$, and $E = -\dfrac{2.0}{1 + x^2}$. Set

$E = -1$ J. This gives a quadratic equation that can be solved for the values of x:

$-1 = -\dfrac{2.0}{1 + x^2} \Rightarrow -1 - x^2 = -2 \Rightarrow x^2 = 1$, which gives $\underline{x =}$

$\underline{\pm 1.0 \text{ m.}}$ The particle will be unbound for $E > 0$.

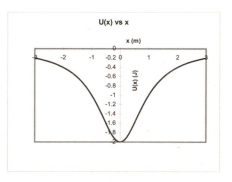

8-33.   The large negative exponents and rapid variation of the potential near $x = 0$ will make this problem difficult to do graphically. However, it can be solved without too much difficulty by using simple algebra and calculus.

(a) The equilibrium point is found by setting the derivative of the potential equal to zero.

$\dfrac{dU}{dx} = -12(1.59 \times 10^{-24})x^{-13} + 6(1.03 \times 10^{-21})x^{-7} = 0$. $x \rightarrow \infty$ satisfies the equation, but this does

not give the minimum energy. Multiply through by x^{13} and 10^{-21} and divide both sides by 6 to get

this equation: $1.03x^6 = 0.00318$, where x is in nm. The result is $x = \left(\dfrac{0.00318}{1.03}\right)^{1/6} = \underline{0.382 \text{ nm.}}$

(b) The lowest energy occurs at the equilibrium point, so substitute the result from (a) into the

equation for U: $U = \dfrac{1.59 \times 10^{-24}}{(0.382)^{12}} - \dfrac{1.03 \times 10^{-21}}{(0.382)^6} = \underline{-1.67 \times 10^{-19} \text{ J.}}$

(c) At the turning points, $K = 0$, so $E = U$. We need to solve this equation:
$-2.0 \times 10^{-21} = (1.59 \times 10^{-24})x^{-12} - (1.03 \times 10^{-21})x^{-6}$. The turning points would be difficult to locate graphically because E is only about 1% of the minimum value of U, which makes it very hard to see on the graph. But the equation isn't as bad as it looks. Multiply through by x^{12} and 10^{21} and rearrange to get $2x^{12} - 1.03x^6 + 0.00159 = 0$. This is a quadratic equation for x^6, and its roots can be used to find the values of x for the turning points:

$$x^6 = \frac{1.03 \pm \sqrt{(1.03)^2 - (8)(0.00159)}}{4} = 0.001548 \text{ nm}^6, \ 0.5135 \text{ nm}^6. \text{ Take the 6th root of each}$$

number to get the values for x: $\underline{x = 0.34 \text{ nm}, 0.89 \text{ nm}}$. (Note: Substituting 0.34 nm in the equation for U gives a result that is wrong. The rapid variation of the potential near $x = 0$ means that rounding x to even three significant digits gives the wrong answer. To get the exact value of U desired, x must be calculated to five significant figures: $x = 0.33979$ nm. However, even this result is suspect because the coefficients in the equation are only given to three significant figures.)

8-35.   Mass of one molecule $= 5m_H + 6m_C + 3m_N + 6m_O = 5(1 \text{ u}) + 6(12 \text{ u}) + 3(14 \text{ u}) + 6(16 \text{ u}) =$

215 u. In kg this is $m = 215 \text{ u} \times \dfrac{1.66 \times 10^{-27} \text{ kg}}{\text{u}} = 3.57 \times 10^{-25} \text{ kg}$.

$E_{molecule} = (4.6 \times 10^6 \text{ J/kg})(3.57 \times 10^{-25} \text{ kg}) = 1.64 \times 10^{-18} \text{ J} \times \dfrac{1 \text{ eV}}{1.60 \times 10^{-19} \text{ J}} = \underline{10.3 \text{ eV/molecule}}$.

8-39. $E = 0.2(150 \text{ kcal}) = 30 \text{ kcal} \times \dfrac{4.187 \times 10^3 \text{ J}}{\text{kcal}} = 1.26 \times 10^4 \text{ J}$. To climb a vertical distance h, this

energy must be converted to potential energy (assuming the person climbs slowly so there's no

significant conversion to kinetic energy). $E = mgh \Rightarrow h = \dfrac{E}{mg}$. Each person should use his or her

own mass in kg to find their personal value for h. Assuming $m = 70$ kg,

$h = \dfrac{1.26 \times 10^4 \text{ J}}{(70 \text{ kg})(9.81 \text{ m/s}^2)} = 183 \text{ m}$.

8-43.   Walking: $t = 0.5$ h. $\underline{E_{walk} = (3.3 \text{ kcal/kg} - \text{h})(0.5 \text{ h}) = 1.7 \text{ kcal/kg}}$.

Slow running plus standing: $t_{run} = \dfrac{2.5 \text{ km}}{8 \text{ km/h}} = 0.313 \text{ h}. \ t_{stand} = 0.5 \text{ h} - 0.313 \text{ h} = 0.187 \text{ h}$.

$\underline{E_{slow} = (8.2 \text{ kcal/kg} - \text{h})(0.313 \text{ h}) + (1.3 \text{ kcal/kg} - \text{h})(0.187 \text{ h}) = 2.8 \text{ kcal/kg}}$.

Fast running plus standing: $t_{run} = \dfrac{2.5 \text{ km}}{16 \text{ km/h}} = 0.156 \text{ h}. \ t_{stand} = 0.5 \text{ h} - 0.156 \text{ h} = 0.344 \text{ h}$.

$\underline{E_{fast} = (15.2 \text{ kcal/kg} - \text{h})(0.156 \text{ h}) + (1.3 \text{ kcal/kg} - \text{h})(0.344 \text{ h}) = 2.8 \text{ kcal/kg}}$. The two running

strategies both consume about the same amount of energy, while walking consumes the least energy.

8-49. $E = m_e c^2 = (9.11 \times 10^{-31} \text{ kg})(9 \times 10^{16} \text{ m}^2/\text{s}^2) \times \dfrac{1 \text{ eV}}{1.60 \times 10^{-19} \text{ J}} \times \dfrac{1 \text{ keV}}{1000 \text{ eV}} = \underline{511 \text{ keV}}$.

$E = m_p c^2 = (1.67 \times 10^{-27} \text{ kg})(9 \times 10^{16} \text{ m}^2/\text{s}^2) \times \dfrac{1 \text{ eV}}{1.60 \times 10^{-19} \text{ J}} \times \dfrac{1 \text{ MeV}}{10^6 \text{ eV}} = \underline{939 \text{ MeV}}$.

8-53. By conservation of energy, the mass energy of the particles produced must be equal to the mass energy of the two original particles plus their kinetic energy. This means $K = 2m_pc^2 - 2m_ec^2 = 2(m_p - m_e)c^2$. The mass of the electron is negligible compared to the mass of the proton, so $K = 2m_pc^2 = 1.88 \times 10^9$ eV (see the solution to 8-45). This is the total kinetic energy of the electron and positron. The kinetic energy of the electron is half of this value, or $K_e = 1.88 \times 10^9$ eV/2 $= 9.40 \times 10^8$ eV.

8-81. (a) $v = 26$ km/h $= 7.22$ m/s. $K = \dfrac{mv^2}{2} = \dfrac{(6.50 \times 10^8 \text{ kg})(7.22 \text{ m/s})^2}{2} = 1.70 \times 10^{10}$ J.

(b) Half of the power output goes toward increasing K, so that power is $P = 22 \times 10^3$ hp

$\times$ 746 W/hp $= 1.64 \times 10^7$ W. Then $\Delta t = \dfrac{K}{P} = \dfrac{1.70 \times 10^{10} \text{ J}}{1.64 \times 10^7 \text{ W}} = 1.03 \times 10^3$ s (17 min).

(c) The time to stop will be the same as the time to go from rest to the final speed, since the magnitude of the change in kinetic energy is the same in both cases. When the ship is slowing

down, its acceleration is $a = \dfrac{v_f - v_i}{\Delta t} = \dfrac{0 - 7.22 \text{ m/s}}{1.03 \times 10^3 \text{ s}} = -7.00 \times 10^{-3}$ m/s^2. (Vector note: the $-$ sign

means the acceleration points in the opposite direction from the initial velocity.) The distance to

stop is given by $v_f^2 - v_i^2 = 2ax \Rightarrow x = \dfrac{v_f^2 - v_i^2}{2a} = \dfrac{0 - (7.22 \text{ m/s})^2}{2(-7.00 \times 10^{-3} \text{ m/s}^2)} = 3.73 \times 10^3$ m (3.73 km,

or 2.3 mi).

8-85. Imagine a cylinder of air with cross sectional area A moving with speed v across the windmill. During a time interval Δt, the length of the cylinder crossing the windmill is $v\Delta t$, and the mass of air crossing the area of the windmill is $\Delta m = \rho A v \Delta t$. The kinetic energy carried by the cylinder of

air is $\Delta K = \dfrac{(\Delta m)v^2}{2} = \dfrac{\rho A v^3 \Delta t}{2}$, so the power delivered to the windmill by the moving air is

$P_{air} = \dfrac{\Delta K}{\Delta t} = \dfrac{\rho A v^3}{2}$. The windmill absorbs 70% of this power, so the power output from the

windmill is $P_w = 0.70 P_{air} = \dfrac{0.70\rho A v^3}{2} = (0.35)(1.29 \text{ kg/m}^3)Av^3 = 0.452 Av^3$. The output of the

generator is 90% of the power absorbed by the windmill, and the motor converts 90% of the generator power to actual useful work. So the power output of the motor is $P_m = (0.90)(0.90)(0.452 Av^3) = 0.366 Av^3$, so if the power output of the motor is specified the

required cross-sectional area is given by $A = \dfrac{P_m}{0.366 v^3}$. $P_m = 2 \times 10^4$ hp $\times \dfrac{746 \text{ W}}{\text{hp}} = 1.49 \times 10^7$ W.

The speed is $v = 40$ km/h $= 11.1$ m/s, so $A = \dfrac{1.49 \times 10^7 \text{ W}}{(0.366)(11.1 \text{ m/s})^3} = 2.98 \times 10^4$ m^2. This

corresponds to a circle with a diameter of $D = \sqrt{\dfrac{4A}{\pi}} = \sqrt{\dfrac{4(1.49 \times 10^4 \text{ m}^2)}{\pi}} = 195$ m.

8-93.   (a)   203.1 km/h = 56.4m/s

Rate gravity does work = Fv

$= (mg \sin \theta)v$

$= 75$ kg $(9.8$ m/s$^2)(\sin 51°)(56.4$ m/s$)$

$= \underline{3.2 \times 10^4}$ J/s

(b) $F_{ice} = \mu N = \mu mg \cos \theta$

$0.03 (75$ kg$) 9.8$ m/s$^2 \cos 51° = 13.9$ N

The power dissipated by ice friction

$= F_{ice}v = (13.9N)(56.5$ m/s$)$

$= \underline{780}$ J/s

Power dissipated by air = Power$_{gravity}$ − Power$_{ice}$

$= 3.1 \times 10^4$ J/s

8-103. $W_I = \int_{0,0}^{1,1} \mathbf{F} \cdot d\mathbf{r} = \int_{0,0}^{1,1} [(4.0\mathbf{i} + 2.0x\mathbf{j}) \cdot (dx\mathbf{i} + dy\mathbf{j})]$. Along path I, $x = y$, so this becomes

$$W_I = \int_{0,0}^{1,1} (4.0\mathbf{i} + 2.0y\mathbf{j}) \cdot (dx\mathbf{i} + dy\mathbf{j}) = \int_{0,0}^{1,1} (4.0dx + 2.0ydy) = \int_0^1 4.0dx + \int_0^1 2.0ydy$$

$= 4.0x\big|_0^1 + y^2\big|_0^1 = 5.0$ J

(b) $W_{II} = \int_{1,1}^{0,1} \mathbf{F} \cdot d\mathbf{r} + \int_{0,1}^{0,0} \mathbf{F} \cdot d\mathbf{r} = \int_{1,1}^{0,1} [(4.0\mathbf{i} + 2.0x\mathbf{j}) \cdot (dx\mathbf{i} + dy\mathbf{j})] + \int_{0,1}^{0,0} [(4.0\mathbf{i} + 2.0x\mathbf{j}) \cdot (dx\mathbf{i} + dy\mathbf{j})]$

From (1,1) to (0,1) y is constant at 1 ($dy = 0$) as x varies from 1 to 0. From (0,1) to (0,0) x is

constant at 0 ($dx = 0$) as y varies from 1 to 0. $W_{II} = \int_1^0 4dx + \int_1^0 0dy = -4.0$ J. If the force were

conservative, we would have $W_{II} = -W_I$. This is not what we find, so we conclude that this force

is not conservative.

8-109.   (a) Since the projectile is traveling horizontally, its potential energy is constant and its total

energy is just equal to its kinetic energy. The instantaneous power is then equal to the

instantaneous rate of change of K: $P = \dfrac{dK}{dt} \dfrac{d}{dt}\left(\dfrac{mv^2}{2}\right) = mv\dfrac{dv}{dt}$. (Don't forget the chain rule!)

$P = (45.36 \text{ kg})(655.9 - 61.1t + 3.26t^2)(-61.1 + 6.52t)$

$= -1.82 \times 10^6 + 3.63 \times 10^5 t - 2.71 \times 10^4 t^2 + 964t^4$

(The magnitude of this is the power "removed from" the projectile.)

(b) At $t = 0$, $K(0) = \dfrac{mv(0)^2}{2} = \dfrac{(45.36 \text{ kg})(655.9 \text{ m/s})^2}{2} = 9.757 \times 10^6$ J. At $t = 3.00$ s,

$K(3.00) = \dfrac{mv(3.00)^2}{2} = \dfrac{(45.36 \text{ kg})[655.9 - (61.1)(3.00) + (3.26)(3.00)^2] \text{ m/s})^2}{2} = \underline{5.71 \times 10^6}$ J.

(c) The average power is $P = \dfrac{\Delta K}{\Delta t} = \dfrac{5.71 \times 10^6 \text{ J} - 9.757 \times 10^6 \text{ J}}{3.00 \text{ s}} = \underline{-1.35 \times 10^6}$ W.

8-111.   The athlete has two legs, each of which can deliver energy at a rate of 200 W. The total energy

change required is equal to the change in potential energy to climb the stairs:

$\Delta E = mgh = (75 \text{ kg})(9.81 \text{ m/s}^2)(457 \text{ m}) = 3.36 \times 10^5$ J. $P = \dfrac{\Delta E}{\Delta t} \Rightarrow \Delta t = \dfrac{\Delta E}{P} = \dfrac{3.36 \times 10^5 \text{ J}}{400 \text{ W}}$

$= \underline{8.4 \times 10^2 \text{ s (14 min)}}$.

8-113.   (a) The total gravitational potential energy stored in the reservoir is $U = mgh = \rho Vgh$. Assuming all the energy can be extracted from the falling water with no kinetic energy left over and no energy lost due to friction, the total energy output will be

$$E = U = (1000 \text{ kg/m}^3)(2.2 \times 10^7 \text{ m}^3)(9.81 \text{ m/s}^2)(270 \text{ m}) \times \frac{1 \text{ kWh}}{3.6 \times 10^6 \text{ J}} = 1.6 \times 10^7 \text{ kWh.}$$

(b) Again assuming all the energy is converted to electrical energy, the power will be

$$P = \frac{dU}{dt} = \frac{d}{dt}(\rho Vgh) = \rho gh \frac{dV}{dt}, \text{ so the volume flow rate will be}$$

$$\frac{dV}{dt} = \frac{P}{\rho gh} = \frac{10^9 \text{ W}}{(1000 \text{ kg/m}^3)(9.81 \text{ m/s}^2)(270 \text{ m})} = 3.8 \times 10^2 \text{ m}^3/\text{s.}$$

9-3.   If an object of mass m is placed at this special point a distance r_1 from the earth and r_2 from the Moon, the net force on it will be zero:

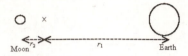

$\dfrac{GM_E m}{r_1^2} - \dfrac{GM_M m}{r_2^2} = 0$. Note that finding the point doesn't depend on the mass m because m will

cancel from the equation. G will also cancel. Let d stand for the Earth–moon distance. Then $r_2 = d$

$- r_1$, and the equation becomes $\dfrac{M_E}{r_1^2} - \dfrac{M_M}{(d-r_1)^2} = 0$, or $r_1^2(M_E - M_m) - 2dM_E r_1 + M_E d^2 = 0$.

Solving the equation in this form will involve working with very large numbers, which increases the likelihood of making a mistake. The numbers can be made more manageable by dividing through by M_E and d^2, and defining a new variable $x = r_1/d$, which gives the location of the equilibrium point as a fraction of d. Doing this gives the following equation:

$x^2(1 - M_m/M_E) - 2x + 1 = 0$. Using the values of the masses from the table in the book finally

gives $0.9877x^2 - 2x + 1 = 0$, which has solutions $x = \dfrac{2 \pm \sqrt{2^2 - 4(0.9877)}}{2(0.9877)} = 1.125,\ 0.900$. The

point must lie between Earth and the Moon, so the correct value of x must be less than 1. That means we reject the first answer and finally get

$r_1 = 0.900d = 0.900(3.84 \times 10^8 \text{ m}) = 3.46 \times 10^8 \text{ m}$.

9-5. $F = \dfrac{GMm}{r^2}$, where M is the mass of the Sun or Moon, $m = 70$ kg, and r is the distance from the

Sun or Moon to the center of Earth. (There are "tidal" effects caused by the fact that objects on one side of Earth are closer to the Sun or Moon than objects on the other side, but those effects will be ignored in this problem.)

For the Sun, $F_{sun} = \dfrac{(6.67 \times 10^{-11} \text{ N} \bullet \text{m}^2/\text{kg}^2)(1.99 \times 10^{30} \text{ kg})(70 \text{ kg})}{(1.50 \times 10^{11} \text{ m})^2} = 0.41$ N

For the Moon, $F_{moon} = \dfrac{(6.67 \times 10^{-11} \text{ N} \bullet \text{m}^2/\text{kg}^2)(7.35 \times 10^{22} \text{ kg})(70 \text{ kg})}{(3.84 \times 10^8 \text{ m})^2} = 2.3 \times 10^{-3}$ N

The force exerted on the person by Earth is $mg = (70 \text{ kg})(9.81 \text{ m/s}^2) = 687$ N. We see that the forces exerted by the Sun and Moon are very small compared to the force exerted by Earth. However, the effect of the forces would be almost unnoticeable anyway, even if they were larger. The reason we notice the force due to Earth is because we are pulled down against the surface, and we feel the reaction force as "weight." If we were in free fall relative to Earth, that reaction force would be absent and we would feel "weightless." The Earth (and everything on it) is in free fall around the Sun and Moon, so there is no sense of "weight" from the attractive forces exerted by those bodies.

9-7. $F = \dfrac{GMm}{r^2}$. The force exerted on the sun by Alpha Centauri is

$$F_{Alpha} = \frac{(6.67 \times 10^{-11} \text{ N} \cdot \text{m}^2/\text{kg}^2)(2.0 \times 10^{30} \text{ kg})(1.99 \times 10^{30} \text{ kg})}{[(4.4 \text{ ly})(9.45 \times 10^{15} \text{ m/ly})]^2} = 1.5 \times 10^{17} \text{ N. The force exerted}$$

on the Sun by Earth is

$$F_{Earth} = \frac{(6.67 \times 10^{-11} \text{ N} \cdot \text{m}^2/\text{kg}^2)(5.98 \times 10^{24} \text{ kg})(1.99 \times 10^{30} \text{ kg})}{(1.50 \times 10^{11} \text{ m})^2} = 3.5 \times 10^{22} \text{ N. The magnitude}$$

of the force exerted by Earth is about five orders of magnitude greater than the force exerted by Alpha Centauri.

9-9. $g = \dfrac{GM}{R^2}$. Use a spreadsheet to calculate the values of g.

Planet	M (kg)	R (m)	g (m/s^2)
Jupiter	1.90E + 27	7.14E + 07	24.9
Saturn	5.67E + 26	6.00E + 07	10.5
Uranus	8.70E + 25	2.54E + 07	8.99

9-13. The 3.0 kg mass is at the origin (0,0). The 4.0 kg mass is located at (0, 2.0 m) and the 7.0 kg mass is at (3.0 m, 0). The net force **F** on the 3.0 kg mass is given by

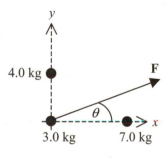

$$\mathbf{F} = \frac{G(3.0 \text{ kg})(7.0 \text{ m})}{(3.0 \text{ m})^2} \mathbf{i} + \frac{G(3.0 \text{ kg})(4.0 \text{ m})}{(2.0 \text{ m})^2} \mathbf{j}$$

$= (1.56 \times 10^{-10} \text{ N})\mathbf{i} + (2.00 \times 10^{-10} \text{ N})\mathbf{j}$. The magnitude of the

force is $F = \sqrt{(1.56)^2 + (2.00)^2} \times 10^{-10} \text{ N} = 2.5 \times 10^{-10}$ N. The

direction is given by $\theta = \tan^{-1} \dfrac{2.00}{1.56} = 52°$. Note that in the

calculation of F the common factor of 10^{-10} was taken out of the square root for simplification, and likewise the common factor of 10^{-10} was omitted in the calculation of θ.

9-15. The magnitude of the force on a mass m is $F = \dfrac{GM_{earth}m}{r^2}$. By Newton's Second Law, $a = \dfrac{F}{m}$, so

$$a = \frac{GM_{earth}}{r^2} = \frac{(6.67 \times 10^{-11} \text{ N} \cdot \text{m}^2/\text{kg}^2)(5.98 \times 10^{24} \text{ kg})}{(3.84 \times 10^8 \text{ m})^2} = 2.71 \times 10^{-3} \text{ m/s}^2. \text{ As a fraction of } g,$$

$$a = \frac{2.71 \times 10^{-3}}{9.81} g = 2.76 \times 10^{-4} g.$$

9-17. Let r stand for the radius of the earth or Io and R stand for the distance from the center of the planet to the center of the moon. Let M stand for the mass of the body causing the tidal force (either the moon or Jupiter), and m stand for the mass of the body affected by the tidal force (either the earth or Io). A diagram is shown for the case of the Earth–Moon system. The difference in centripetal accelerations from the near side to far side of the affected body is

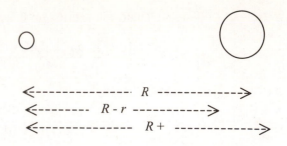

$$\Delta a = GM\left[\frac{1}{(R-r)^2} - \frac{1}{(R+r)^2}\right] = GM\left[\frac{(R+r)^2 - (R+r)^2}{(R-r)^2(R+r)^2}\right] = GM\left[\frac{4rR}{(R-r)^2(R+r)^2}\right]. \text{ Divide}$$

through by R^4 to change the variable to r/R:

$$\Delta a = GM\left[\frac{(R+r)^2 - (R+r)^2}{(R-r)^2(R+r)^2}\right] = \frac{GM}{R^2}\left[\frac{4(r/R)}{(1-r/R)^2(1+r/R)^2}\right]. \text{ Note that one factor of } R^2 \text{ was}$$

brought out under GM while the other factor was put under rR in the numerator.

For the Earth–Moon system, $r/R = \dfrac{6.378 \times 10^6 \text{ m}}{3.84 \times 10^8 \text{ m}} = 0.0166$, and at the orbit of the Moon

$$\frac{GM}{R^2} = \frac{(6.67 \times 10^{-11} \text{ N} \cdot \text{m}^2/\text{kg}^2)(7.35 \times 10^{22} \text{ kg})}{(3.84 \times 10^8 \text{ m})^2} = 3.33 \times 10^{-5} \text{ m/s}^2. \text{ Thus}$$

$$\Delta a = (3.33 \times 10^{-5} \text{ m/s}^2)\left[\frac{4(0.0166)}{(0.9834)^2(1.0166)^2}\right] = 2.21 \times 10^{-7} \text{ m/s}^2. \text{ The surface gravity on Earth is}$$

$g = 9.81 \text{ m/s}^2$, so $\dfrac{\Delta a}{g} = 2.25 \times 10^{-7}$. For the Jupiter–Io system, $r/R = \dfrac{1.82 \times 10^6 \text{ m}}{4.22 \times 10^8 \text{ m}} = 4.31 \times 10^{-3}$

and $\dfrac{GM}{R^2} = \dfrac{(6.67 \times 10^{-11} \text{ N} \cdot \text{m}^2/\text{kg}^2)(1.90 \times 10^{27} \text{ kg})}{(4.22 \times 10^8 \text{ m})^2} = 0.712 \text{ m/s}^2.$

$$\Delta a = (0.712 \text{ m/s}^2)\left[\frac{4(4.31 \times 10^{-3})}{(0.99569)^2(1.00431)^2}\right] = 0.0123 \text{ m/s}^2. \text{ The surface gravity on Io is}$$

$$g = \frac{Gm}{r^2} = \frac{(6.67 \times 10^{-11} \text{ N} \cdot \text{m}^2/\text{kg}^2)(8.9 \times 10^{22} \text{ kg})}{(1.82 \times 10^6 \text{ m})^2} = 1.79 \text{ m/s}^2, \text{ which gives } \frac{\Delta a}{g} = 6.86 \times 10^{-3}.$$

We see that the effects (absolute and relative) caused by Jupiter on Io are much larger than the effects caused by the Moon on Earth.

9-25. $T^2 = \dfrac{4\pi^2 r^3}{GM_J} \Rightarrow T = \sqrt{\dfrac{4\pi^2 r^3}{GM_J}} = \sqrt{\dfrac{4\pi^2}{(6.67 \times 10^{-11} \text{ N} \cdot \text{m}^2/\text{kg}^2)(1.90 \times 10^{27} \text{ kg})}} r^{3/2},$ or

$T = 1.765 \times 10^{-8} r^{3/2}$, where r is in m and T is in s. Use a spreadsheet to do the calculations.

Moon	r (m)	T (s)	T (days)
Io	4.22E + 08	1.53E + 05	1.77
Europa	6.71E + 08	3.07E + 05	3.55
Ganymede	1.07E + 09	6.18E + 05	7.15

9-35. From Problem 9-34, $\dfrac{r_1}{r_2} = \dfrac{10M_S}{25M_S} \Rightarrow r_1 = \dfrac{2}{5}r_2$. d is the

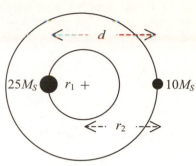

separation between the stars, so $d = r_1 + r_2 = \dfrac{7}{5}r_2$. The

gravitational attraction between the stars provides the centripetal force that keeps them in orbit:

$F = \dfrac{G(25M_S)(10M_S)}{d^2}$. For the $10M_S$ black hole orbiting at

a distance r_2 with a period T, we get

$\dfrac{G(25M_S)(10M_S)}{d^2} = \dfrac{(10M_S)v^2}{r_2} = \dfrac{(10M_S)(4\pi^2 r_2)}{T^2}$. Substituting d in terms of r_2 and solving for r_2

gives $r_2 = \left[\dfrac{625GM_S T^2}{196\pi^2} \right]^{1/3}$. The period is $T = 5.6$ days $= 4.84 \times 10^5$ s. The final result is

$r_2 = \left[\dfrac{625(6.67 \times 10^{-11}\ \text{N} \cdot \text{m}^2/\text{kg}^2)(1.99 \times 10^{30}\ \text{kg})(4.84 \times 10^5\ \text{s})^2}{196\pi^2} \right]^{1/3} = 2.16 \times 10^{10}$ m. The

separation is $d = \dfrac{7}{5}r_2 = 3.0 \times 10^{10}$ m.

9-37. (a) "Low altitude" means that the radius of the orbit is essentially the same as the radius of the

earth, 6.38×10^6 m. The period of a satellite in such an orbit is $T = \sqrt{\dfrac{4\pi^2 r^3}{GM}} =$

$\sqrt{\dfrac{4\pi^2 (6.38 \times 10^6\ \text{m})^3}{(6.67 \times 10^{-11}\ \text{N} \cdot \text{m}^2/\text{kg}^2)(5.98 \times 10^{24}\ \text{kg})}} = 5.07 \times 10^3$ s, so its orbital speed is $v_{orbit} = \dfrac{2\pi r}{T} =$

$\dfrac{2\pi(6.38 \times 10^6\ \text{m/s})}{5.07 \times 10^3\ \text{s}} = 7.91 \times 10^3$ m/s. A point on the surface of the earth makes one complete

rotation every 24 hours, so the tangential speed of a point at latitude 28° is

$v_{tan} = \dfrac{2\pi(6.38 \times 10^6\ \text{m})\cos 28°}{(24\ \text{h})(3600\ \text{s/h})} = 410$ m/s. (Note that the tangential speed is much less than the

orbital speed! Objects on the surface of the earth are not in orbit.) The earth rotates from west to east, so if a satellite is launched toward the east, it gains 410 m/s from the earth, and the required launch speed is reduced by that amount: $v_{launch} = 7.91 \times 10^3$ m/s $- 410$ m/s $= 7.50 \times 10^3$ m/s. . If

the satellite is launched toward the west, its launch speed must be increased:

$v_{launch} = 8.32 \times 10^3$ m/s.

(b) Launching east, $K = \dfrac{mv_{launch}^2}{2} = \dfrac{(14.0\ \text{kg})(7.50 \times 10^3\ \text{m/s})^2}{2} = 3.94 \times 10^8$ J. Launching west,

$K = \dfrac{mv_{launch}^2}{2} = \dfrac{(14.0\ \text{kg})(8.32 \times 10^3\ \text{m/s})^2}{2} = 4.85 \times 10^8$ J.

9-41. Even though it's not explicitly given in the text, the mathematical statement of Kepler's Third Law for elliptical orbits is $T = \sqrt{\dfrac{4\pi^2 a^3}{GM_S}}$, where a is the semimajor axis. For Sputnik I, $a = 6.97 \times 10^6$ m, and $T = \sqrt{\dfrac{4\pi^2 (6.97 \times 10^6 \text{ m})^3}{(6.67 \times 10^{-11} \text{ N}\bullet\text{m}^2/\text{kg}^2)(5.98 \times 10^{24} \text{ kg})}} = \underline{5.79 \times 10^3 \text{s} \ (96.5 \text{ min})}$. For

Explorer I, $a = 7.83 \times 10^6$ m, and $T = \sqrt{\dfrac{4\pi^2 (7.83 \times 10^6 \text{ m})^3}{(6.67 \times 10^{-11} \text{ N}\bullet\text{m}^2/\text{kg}^2)(5.98 \times 10^{24} \text{ kg})}} = $

$\underline{6.89 \times 10^3 \text{ s} \ (115 \text{ min})}$.

9-43. Using the equation given for Problem 9-41, $a = \left(\dfrac{GM_S T^2}{4\pi^2}\right)^{1/3}$, where

$T = 2380$ yr $= 7.326 \times 10^{10}$ s.

$a = \left[\dfrac{(6.67 \times 10^{-11} \text{ N}\bullet\text{m}^2/\text{kg}^2)(1.99 \times 10^{30} \text{ kg})(7.326 \times 10^{10} \text{ s})^2}{4\pi^2}\right]^{1/3}$

$= 2.67 \times 10^{13}$ m. From the diagram 9-42, we see that

$r_{min} + r_{max} = 2a \Rightarrow r_{max} = 2a - r_{min}$

$= 2(2.67 \times 10^{13} \text{ m}) - 1.37 \times 10^{11} \text{ m} = 5.33 \times 10^{13}$ m, or $\underline{5.33 \times 10^{10} \text{ km}}$. This is about 10 times Pluto's mean orbital distance from the Sun.

9-51. For an escape speed of c, $v_{exc} = c = \sqrt{\dfrac{2GM_{earth}}{R}} \Rightarrow R = \dfrac{2GM_{earth}}{c^2}$. Substituting numerical values

gives $R = \dfrac{2(6.67 \times 10^{-11} \text{ N}\bullet\text{m}^2/\text{kg}^2)(5.98 \times 10^{24} \text{ kg})}{(3 \times 10^8 \text{m/s})^2} = 8.86 \times 10^{-3}$ m, or $\underline{8.86 \text{ mm}}$.

9-53. $v_{esc} = \sqrt{\dfrac{2GM_{moon}}{R_{moon}}} = \sqrt{\dfrac{2(6.67 \times 10^{-11} \text{ N}\bullet\text{m}^2/\text{kg}^2)(7.35 \times 10^{22} \text{ kg})}{1.74 \times 10^6 \text{ m}}} = \underline{2.37 \times 10^3 \text{ m/s}}.$

$\dfrac{600}{2.37 \times 10^3} = 0.253$, so the average speed is about 25% of the escape speed. But since some

molecules are moving faster than 600 m/s, some of them will be moving fast enough to escape into space. Over a long period of time, essentially all of the gas molecules will eventually be lost.

9-55. (a) Conservation of energy says $K_{earth} + U_{earth} = K_{moon} + U_{moon}$. If we take the kinetic energy at the moon to be essentially zero and neglect the potential energy due to the moon's gravity, then the muzzle speed of the cannon must satisfy $\dfrac{mv^2}{2} - \dfrac{GM_{earth}m}{R_{earth}} = -\dfrac{GM_{earth}m}{d}$, where d is the earth–

moon distance. This gives $v = \sqrt{2GM_{earth}\left(\dfrac{1}{R_{earth}} - \dfrac{1}{d}\right)}$

$= \sqrt{2(6.67 \times 10^{-11} \text{ N}\bullet\text{m}^2/\text{kg}^2)(5.98 \times 10^{24} \text{ kg})\left(\dfrac{1}{6.38 \times 10^6 \text{ m}} - \dfrac{1}{3.84 \times 10^8 \text{ m}}\right)} = \underline{1.11 \times 10^4 \text{ m/s}}.$

(b) $K = \dfrac{mv^2}{2} = \dfrac{(2000 \text{ kg})(1.11 \times 10^4 \text{ m/s})^2}{2} = 1.23 \times 10^{11}$ J. The mass of TNT required is

$m = \dfrac{1.23 \times 10^{11} \text{ J}}{4.2 \times 10^9 \text{ J/ton}} = \underline{29 \text{ tons.}}$

(c) $v^2 = 2ax \Rightarrow a = \dfrac{v^2}{2x} = \dfrac{(1.11 \times 10^4 \text{ m/s})^2}{2(500 \text{ m})}. = 1.23 \times 10^5 \text{ m/s}^2.$

9-63.   (a) 140 km/hr = 38.9 m/s. For a stable orbit, the centripetal acceleration is provided by the gravitational attraction, or $F_c = mv^2/R = GMm/R^2$.

The speed required for a stable orbit is therefore given by $v^2 = GM/R$, where $R = 3.76 \times 10^{19}$ kg, and $R = 195 \times 10^3$ m.

Then $v = 113.4$ m/s, so that the ball will not orbit around Mimas.

(b) The ball will rise until its initial kinetic energy, $(1/2)mv^2$, is equal to the change in potential energy,

$U_{\text{final}} - U_{\text{initial}} = -GMm/(R + h) + GMm/R.$

Then $(1/2)mv^2 = GMm[1/R - 1/(R + h)]$

solving for h, $v^2R^2 + v^2Rh = 2GMh$, from which $\underline{h = 5910 \text{ m}}$

9-67.   In Problem 9-43, we found the semimajor axis is related to perihelion and aphelion distances by

$a = \dfrac{r_{min} + r_{max}}{2}$, so $a = \dfrac{5.06 + 61.25}{2} \times 10^7$ km $= 3.32 \times 10^{11}$ m. In terms of a, the total energy

is $E = -\dfrac{GM_S m}{2a}$. At perihelion, $E = -\dfrac{GM_S m}{2a} = \dfrac{mv_{max}^2}{2} - \dfrac{GM_S m}{r_{min}}$, so

$v_{max} = \sqrt{GM_S \left(\dfrac{2}{r_{min}} - \dfrac{1}{a} \right)}$

$= \sqrt{(6.67 \times 10^{-11} \text{ N} \bullet \text{m}^2/\text{kg}^2)(1.99 \times 10^{30} \text{ kg}) \left(\dfrac{2}{5.06 \times 10^{10} \text{ m}} - \dfrac{1}{3.32 \times 10^{11} \text{ m}} \right)}$

$= \underline{6.96 \times 10^4 \text{ m/s}}$. To find the speed v_{min} at aphelion, we could use the energy equation again, but

it's simpler to use Kepler's Second Law: $r_{min}v_{max} = r_{max}v_{min} \Rightarrow v_{min} = \dfrac{r_{min}v_{max}}{r_{max}}$, which gives

$v_{min} = \dfrac{(5.06)(6.96 \times 10^4 \text{ m/s})}{61.25} = \underline{5.75 \times 10^3 \text{ m/s}}$. Note that it was not necessary to include the

units or the powers of 10 for the perihelion and aphelion distances, because they cancel in the ratio.

9-69.   In the solution to Problem 9-50, it was stated without proof that the energy of a body in an elliptical orbit depends on the semimajor axis. In this problem, we'll prove that statement. In Problem 9-43 we found that $2a = r_1 + r_2$ just by looking at the geometry of an ellipse, so the expression for the energy in this problem is the same as the statement in solution 9-50. In Problem 9-68 it was shown that energy conservation in an elliptical orbit around the sun means

$\dfrac{v_1^2}{2} - \dfrac{GM_S}{r_1} = \dfrac{v_2^2}{2} - \dfrac{GM_S}{r_2}$. Kepler's Second Law says $r_1v_1 = r_2v_2 \Rightarrow v_2 = \dfrac{r_1v_1}{r_2}$. Substituting this

into the energy equation gives $\dfrac{v_1^2}{2} - \dfrac{GM_E}{r_1} = \dfrac{v_1^2 r_1^2}{2r_2^2} - \dfrac{GM_E}{r_2} \Rightarrow v_1^2 \left(1 - \dfrac{r_1^2}{r_2^2} \right) = 2GM_S \left(\dfrac{1}{r_1} - \dfrac{1}{r_2} \right).$

Solving for v_1 gives $v_1^2 = \dfrac{2GM_S\left(\dfrac{1}{r_1} - \dfrac{1}{r_2}\right)}{\left(1 - \dfrac{r_1^2}{r_2^2}\right)} = \dfrac{2GM_S\left(\dfrac{r_2 - r_1}{r_1 r_2}\right)}{\left(1 - \dfrac{r_1}{r_2}\right)\left(1 + \dfrac{r_1}{r_2}\right)} = \dfrac{2GM_S r_2 (r_2 - r_1)}{r_1 (r_2 - r_1)(r_2 + r_1)} = \dfrac{2G r_2 M_S}{r_1 (r_2 + r_1)}.$

Then $\dfrac{E}{m} = \dfrac{v_1^2}{2} - \dfrac{GM_S}{r_1} = \dfrac{G r_2 M_S}{r_1 (r_2 + r_1)} - \dfrac{GM_S}{r_1} = GM_S\left[\dfrac{r_2 - (r_2 + r_1)}{r_1 (r_2 + r_1)}\right] \Rightarrow \underline{E = -\dfrac{GM_S}{r_2 + r_1}}.$ So the

statement made in 9-50 is true.

9-71. The maximum distance from the center of the Moon is labeled r in the
sketch of the orbit. Using conservation of energy,
$\dfrac{mv_{launch}^2}{2} - \dfrac{GM_M m}{R_M} = \dfrac{mv^2}{2} - \dfrac{GM_M m}{r}$, where v is the speed of the

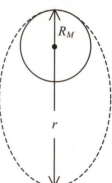

projectile at r. According to Kepler's Second Law, $R_M v_{launch} = rv$.
Because r is in the denominator in the energy equation, it will be easier
to find v and then solve for r rather than vice versa. Cancel the mass m of
the satellite from the energy equation and use the second equation to get
r in terms of v. Substitute that expression into the energy equation to get
$\dfrac{v_{launch}^2}{2} - \dfrac{GM_M}{R_M} = \dfrac{v^2}{2} - \dfrac{GM_M v}{R_M v_{launch}}$. Multiply through by 2 and rearrange to get a quadratic equation

for v: $v^2 - \left(\dfrac{2GM_M}{R_M v_{launch}}\right)v + \left(\dfrac{2GM_M}{R_m} - v_{launch}^2\right) = 0$. Substituting numerical data and using $v_{launch} = $

2000 m/s gives $v^2 - 2818v + 1.635 \times 10^6 = 0$. The roots are

$v = \dfrac{2828 \pm \sqrt{2818^2 - 4(1.635 \times 10^6)}}{2} = 2000$ m/s, 817 m/s. The first root is the launch speed, so

the speed at r is $\underline{v = 817 \text{ m/s}}$.

From Kepler's Second Law, $r = \dfrac{R_M v_{launch}}{v} = 4.26 \times 10^6$ m. The problem asks for the height, so R_M

must be subtracted from this radius: $\underline{h = 2.52 \times 10^6 \text{ m}}$.

9-85. At the lowest point ($r_1 = 6378$ km $+ 200$ km), $E = \dfrac{mv^2}{2} - \dfrac{GM_E m}{r_1}$. At the highest point, $K = 0$

and $E = -\dfrac{GM_E m}{r_2}$. By conservation of energy, $-\dfrac{GM_E m}{r_2} = \dfrac{mv^2}{2} - \dfrac{GM_E m}{r_1}$, which gives

$r_2 = \left(\dfrac{1}{r_1} - \dfrac{v^2}{2GM_E}\right)^{-1} = \left(\dfrac{1}{6.578 \times 10^6 \text{ m}} - \dfrac{(8.50 \times V10^3 \text{ m/s})^2}{2(6.67 \times 10^{-11} \text{ N} \cdot \text{m}^2/\text{kg}^2)(5.98 \times 10^{24} \text{ kg})}\right)^{-1}$

$= 1.626 \times 10^7$ m

The height above the earth is $\underline{h_2 = r_2 - R_E = 1.626 \times 10^7 \text{ m} - 6.378 \times 10^6 \text{ m} = 9.89 \times 10^6 \text{ m}}.$

9-87. (a) For a circular orbit, $E = -\dfrac{GM_E m}{2r} = -\dfrac{(6.67 \times 10^{-11} \text{ N} \cdot \text{m}^2/\text{kg}^2)(5.98 \times 10^{24} \text{ kg})(700 \text{ kg})}{2(4.23 \times 10^7 \text{ m})}$

$= \underline{-3.30 \times 10^9 \text{ J}}.$

(b) For a parabolic orbit, $E = 0$, so the satellite must be given $\underline{3.30 \times 10^9 \text{ J}}$ of extra energy for it to
escape.

CHAPTER 10   SYSTEMS OF PARTICLES

10-1.   No directions are given in the problem, so we'll only calculate magnitudes of momentum.

$p_{bullet} = m_{bullet}v_{bullet} = (0.015 \text{ kg})(600 \text{ m/s}) = 9.0 \text{ kg} \cdot \text{m/s}.$

$p_{arrow} = m_{arrow}v_{arrow} = (0.040 \text{ kg})(80 \text{ m/s}) = 3.2 \text{ kg} \cdot \text{m/s}.$

10-3.   No directions are given in the problem, so we'll only calculate magnitudes of momentum. A good way to approach this kind of problem is to use a spreadsheet. Fill in the information from the tables in Chapters 1 and 2 and let the spreadsheet do the calculations.

Object	m (kg)	v (m/s)	p (kg-m/s)
Earth	6.0E+24	3.0E+04	1.8E+29
Jet	1.6E+05	2.7E+02	4.3E+07
Car	1.5E+03	25	3.8E+04
Man	73	1.3	95
Electron	9.1E-31	2.2E+06	2.0E-24

10-5. $\mathbf{p} = mv\cos 20°\mathbf{i} + mv\sin 20°\mathbf{j}$

$= (9.1 \times 10^{-31} \text{ kg})(2.0 \times 10^5 \text{ m/s})(\cos 20°\mathbf{i} + \sin 20°\mathbf{j})$

$= (1.6 \times 10^{-25}\mathbf{i} + 7.7 \times 10^{-26}\mathbf{j})\text{kg} \cdot \text{m/s}.$

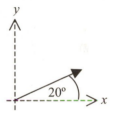

10-7.   (a) $\mathbf{p}_i = m\mathbf{v}_i = (0.43 \text{ kg})(26 \text{ m/s})$ @ 30° upward $= 11.2 \text{ kg} \cdot \text{m/s}$ upward.

Defining x and y directions as shown, this is

$\mathbf{p}_i = 11.2\cos 30°\mathbf{i} + 11.2\sin 30°\mathbf{j} = 9.7\mathbf{i} + 5.6\mathbf{j} \text{ kg} \cdot \text{m/s}.$

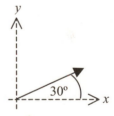

(b) If there's no air resistance, the horizontal component of velocity, and hence the horizontal component of momentum, remains constant. Thus at the highest point in the trajectory, the vertical velocity and therefore the vertical momentum are zero, and $\mathbf{p}_{highest} = 9.7\mathbf{i} + 0\mathbf{j} \text{ kg} \cdot \text{m/s}.$

(c) When the ball reaches the ground again, its horizontal component of velocity is still unchanged but its vertical component of velocity points down. Therefore the ball's momentum is $\mathbf{p}_f = 9.7\mathbf{i} - 5.6\mathbf{j} \text{ kg} \cdot \text{m/s}.$ The final momentum is $\underline{\text{not}}$ the same as the initial momentum!

10-19.   The projectile velocity is $\mathbf{v}_1 = v_1 \cos 20°\mathbf{i} + v_1 \sin 20°\mathbf{j}$. In the y direction the total mass that recoils is the mass of the cannon plus the mass of the entire earth, since the cannon pushed down on the earth. Because the mass of the earth is so huge (6×10^{24} kg), the y component of the recoil velocity is unmeasurably small. In the x direction, only the cannon recoils. so the x component of the recoil velocity is not zero. The total velocity is

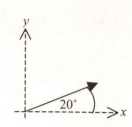

$$\mathbf{v}'_2 = -\frac{m_1}{m_2}v_1 \cos 20°\mathbf{i} - \frac{m_1}{m_2 + m_E}v_1 \sin 20°\mathbf{j} = -\frac{6}{2200}(500 \text{ m/s})(\cos 20°)\mathbf{i} + 0\mathbf{j} = \underline{-(1.3 \text{ m/s})\mathbf{i} + 0\mathbf{j}}.$$

10-23. $F_x = \frac{dp_x}{dt} \Rightarrow p_{x,2} - p_{x,1} = \Delta p_x = \int_{t_1}^{t_2} F_x dt$. Substitute the given expression for F_x:

$$\Delta p_x = \int_0^{5.0 \text{ s}}(2.0t + 3.0t^2)dt = (t^2 + t^3)\Big|_0^{5.0 \text{ s}} = 150 \text{ N·s}.$$ Note that the coefficients 2.0 and 3.0 have

units of N/s and N/s^2, respectively, so the two terms in the result of the integration are actually multiplied by 1 N/s and 1 N/s^2, respectively. Also note that the unit N • s is the same as kg • m/s.

10-25.   Assuming the nucleus is initially at rest, the initial momentum is zero. Conservation of momentum requires $\mathbf{p}_N + \mathbf{p}_e + \mathbf{p}_\nu = 0$. Label the components of the nucleus's momentum as $p_{N,x} = m_N v_x$, $p_{N,x} = m_N v_x$. Using these and resolving p_ν into components gives $(m_N v_x + p_e + p_\nu \cos 30°)\mathbf{i} + (m_N v_y + p_\nu \sin 30°)\mathbf{j} = 0$. Setting

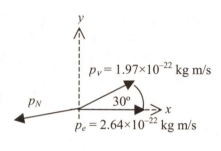

each component equal to zero gives $v_x = -\dfrac{p_e + p_\nu \cos 30°}{m_N}$,

$$v_x = -\frac{p_\nu \sin 30°}{m_N}.$$

$m_N = (63.9 \text{ u})(1.66 \times 10^{-27} \text{ kg/u}) = 1.06 \times 10^{-25}$ kg. Substituting this and the other numerical values in the problem gives $v_x = -4.10 \times 10^3$ m/s, $v_y = -929$ m/s. The minus signs mean the components point in the $-x$ and $-y$ directions, respectively. The recoil velocity in component form is $\mathbf{v} = \underline{-(4.10 \times 10^3\mathbf{i} + 929\mathbf{j}) \text{ m/s}}$. As a magnitude and direction, the recoil velocity is

$$v = \sqrt{(4.10 \times 10^3)^2 + (929)^2} \text{ m/s}, \quad \theta = \tan^{-1}\frac{929}{1.04 \times 10^3} = 12.8°.$$ Because both components of $\mathbf{v}$ are

negative, θ is in the third quadrant. To get the correct angle measured from the $+x$ direction, we must add 180° to the value from the calculator, which gives $\underline{\theta = 193°}$. Another possibility is to say that $\theta = 12.8°$ _below_ the $-x$ axis.

10-33.   The woman (m_1) is at $x = 0$ and the man (m_2) is at $x = 3.5$ m. The location of the center of mass is given by

$$x_{CM} = \frac{\sum m_i x_i}{\sum m_i} = \frac{(39 \text{ kg})(0) + (72 \text{ kg})(3.5 \text{ m})}{111 \text{ kg}} = \underline{1.9 \text{ m}}.$$

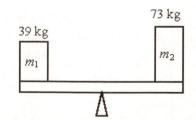

10-35. $x_{CM} = \dfrac{M_J R_J}{M_S + M_J} = \dfrac{(1.90 \times 10^{27} \text{ kg})(7.78 \times 10^8 \text{ m})}{1.99 \times 10^{30} \text{ kg}}$

$= \underline{7.42 \times 10^5 \text{ m}}$. Note that to three significant figures, the mass of Jupiter is negligible compared to the mass of the Sun. The distance to the CM is $\dfrac{7.42 \times 10^5}{6.96 \times 10^8} = 1.07 \times 10^{-3}$,

or 0.107% of the sun's radius.

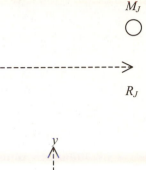
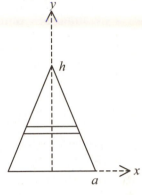

10-37. Assume the base of the triangle has its ends at $x = \pm a$, and the apex is at $y = h$. By symmetry, $\underline{x_{CM} = 0}$. The right side of the triangle is defined by the line $y = -\dfrac{h}{a}x + h$. Imagine a strip across the triangle a distance y up from the base with a length $2x$ and a thickness dy as shown. The total area of the triangle is $A = \dfrac{\text{base} \times \text{height}}{2} = ah$,

since the base has a length of $2a$. If the mass M in the triangle is uniformly distributed, then the mass dm in the strip is $dm = \dfrac{M}{A}(2xdy) = \dfrac{2Mxdy}{ah}$. Inverting the equation for $y(x)$ and substituting in the equation for dm gives

$dm = \dfrac{2M\left(a - \dfrac{a}{h}y\right)dy}{ah} = \dfrac{2M}{h}\left(1 - \dfrac{y}{h}\right)dy$. The y coordinate of the center of mass is

$y_{CM} = \dfrac{\displaystyle\int_{y=0}^{y=h} ydm}{M} = \dfrac{2}{h}\int_0^h y\left(1 - \dfrac{y}{h}\right)dy = \dfrac{2}{h}\int_0^h \left(y - \dfrac{y^2}{h}\right)dy = \dfrac{2}{h}\left(\dfrac{y^2}{2} - \dfrac{y^3}{3h}\right)\Bigg|_0^h = \dfrac{2}{h}\left(\dfrac{h^2}{2} - \dfrac{h^2}{3}\right) = \underline{\dfrac{h}{3}}$.

10-45. Let the axes be as shown. The center of masses of the three sheets are at $(L/2, L/2, 0)$, $(L/2, 0, L/2)$ and $(0, L/2, L/2)$. Let the mass of each sheet be M. Then center of mass of entire structure is:

$m(L/2, L/2, 0) + m(L/2, 0, L/2) + m(0, L/2, L/2)/3M$

$= (L, L, L)/3 = \underline{(L/3, L/3, L/3)}$

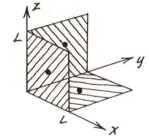

10-49.   As in Example 6, take the y axis to point down with the origin at the apex of the cone. Call the vertex angle ϕ. Then the circular slab at a distance y from the vertex has a radius $r = y \tan\phi$ and a thickness dy. The mass of the slab is $dm = \rho(\pi r^2)dy = \pi\rho y^2 \tan^2\phi dy$. Then y_{CM} is given by

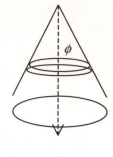

$$y_{CM} = \frac{\int_{y=0}^{y=h} y\,dm}{\int dm} = \frac{\int_0^h \pi\rho y^3 \tan^2\phi dy}{\int_0^h \pi\rho y^2 \tan^2\phi dy} = \frac{\int_0^h y^3 dy}{\int_0^h y^2 dy} = \frac{3h}{4}, \text{ where } h \text{ is the}$$

height of the cone. Since y points down, this result means the center of mass is $h/4$ from the bottom of the cone. (This is the same result for the pyramids in Example 6 and Problem 38!) For Mount Fuji, $h = 3800$ m, so the center of mass is <u>950 m from the base of the volcano</u>.

10-51.   Do this by finding the change in height of the CM of the arm—ignore the rest of the body. When the arms are hanging at the side of the body, its CM is located at

$$h_1 = \frac{(0.066\text{ M})(0.717\text{ L}) + (0.042\text{ M})(0.553\text{ L}) + (0.017\text{ M})(0.431\text{ L})}{(0.066\text{ M} + 0.042\text{ M} + 0.017\text{ M})} = 0.623\text{ L. When the arms}$$

are held horizontally, all the segments are at shoulder height $h_2 = 0.812$ L. The change in potential energy is $\Delta U = mg(h_2 - h_1) = mg(0.812 - 0.623)\text{L} = 0.189\ mgL$. The total mass lifted is $m = 0.066\text{ M} + 0.042\text{ M} + 0.017\text{ M} = 0.125\text{ M} \Rightarrow \Delta U = 0.0236\ MgL$. Substituting the numerical data gives <u>$\Delta U = 0.0236(80\text{ kg})(9.81\text{ m/s}^2)(1.70\text{ m}) = 31.5$ J</u>.

If the arms are now lifted to a vertical position, the CMs are lifted to heights above the shoulder that are the same as their initial distances below the shoulder. Thus ΔU in lifting the arms from the horizontal position to the vertical position is the same as for lifting the arms from the lowered position to the horizontal position, or 31.5 J. The change in potential energy to lift the arms from the lowered position to the vertical position will be twice this: <u>$\Delta U = 63$ J</u>.

10-55.   The diagram shows a cross section through the shell. Take the x axis to point left as shown. By symmetry, the CM must lie on the x axis. Imagine that the shell is divided in rings. At a distance x from the origin, the radius of a ring will be $r = R\cos\theta$, and $x = R\sin\theta$. If we imagine a ring with thickness $Rd\theta$, then the mass in the ring is

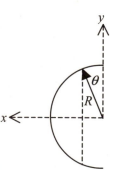

$$dm = \frac{M}{2\pi R^2}dA = \frac{M}{2\pi R^2}2\pi r(Rd\theta) = \frac{Mr}{R}d\theta = M\sin\theta d\theta, \text{ where we have}$$

used the area of a hemisphere to find the mass per unit area. Now

$$x_{CM} = \frac{\int_{x=0}^{x=R} x\,dm}{M} = R\int_0^{\pi/2} \sin\theta\cos\theta d\theta = \frac{R\sin^2\theta}{2}\Big|_0^{\pi/2} = \frac{R}{2}, \text{ where } x \text{ was given}$$

in terms of θ, so going from $x = 0$ to $x = R$ is equivalent to going from $\theta = 0$ to $\theta = \pi/2$.

10-57. According to Eq. 10.37, the velocity of the CM is related to the particle velocities by

$$\mathbf{v}_{CM} = \frac{\sum_i m_i \mathbf{v}_i}{\sum_i m_i}.$$ For two identical particles with one having zero velocity, $\mathbf{v}_{CM} = \frac{m_P \mathbf{v}}{2m_P} = \frac{\mathbf{v}}{2}.$

The speed of the moving proton is given by $v = \sqrt{\dfrac{2K}{m_P}} = \sqrt{\dfrac{2(1.6 \times 10^{-13} \text{ J})}{1.67 \times 10^{-27} \text{ kg}}} = 1.38 \times 10^7$ m/s. The

speed of the CM will be half of this, and the direction will be in the same direction as the velocity of the moving proton. If we assume the proton is moving in the $+x$ direction, then $\mathbf{v}_{CM} = 6.9 \times 10^6 \mathbf{i}$ m/s.

10-59. Since there are no external forces, the position of the CM remains unchanged. Let the CM be at the origin. This is where the boat finally meets the shark. Therefore the original positions of the shark and the boat are

$x_{sh} = 300 - 45 = 255$ m

$x_b = -45$ m

Thus,

$$\frac{m_{sh} x_{sh} + m_b x_b}{m_{sh} + m_b} = (m_{sh}\,255 + 5400(-45))/(m_{sh} + 5400) = x_{CM} = 0$$

which gives

$$m_{sh} = \frac{5400 \times 45}{255} = 953 \text{ kg}$$

10-67. To solve this problem, we begin with the spreadsheet created for Problem 42. If we assume that the orbits are essentially circular, then the orbital speed of a planet is $v = \dfrac{2\pi r}{T}$, where r is the mean radius of the orbit and T is the period of the orbit. We also know that the velocity of each planet is perpendicular to its radius vector from the sun. The motion of the planets is counterclockwise, so the direction of the velocity for a planet is equal to its position angle plus $90°$. In component form, the velocity is $\mathbf{v} = v\cos(\theta + 90°)\mathbf{i} + v\sin(\theta + 90°)\mathbf{j}$. The information is tabulated in the spreadsheet.

Planet	m (kg)	r (km)	$\theta(°)$	T (yr)	v (km/s)	v_x (km/s)	v_y (km/s)
Mercury	3.30E + 23	5.79E + 07	232.0	0.241	47.77	37.643	−29.410
Venus	4.87E + 24	1.08E + 08	170.0	0.615	34.92	−6.063	−34.387
Earth	5.98E + 24	1.50E + 08	95.5	1.00	29.83	−29.688	−2.859

$$v_{CM,x} = \frac{(0.33)(37.64) + (4.87)(-6.063) + (5.98)(-26.69)}{0.33 + 4.87 + 5.98} = -17.4 \text{ km/s}$$

$$v_{CM,y} = \frac{(0.33)(-29.41) + (4.87)(-34.39) + (5.98)(-2.859)}{0.33 + 4.87 + 5.98} = -17.4 \text{ km/s}$$

Note that it was not necessary to carry along the powers of 10 for the masses, since they cancel. These equations give the x and y components of $\mathbf{v}_{CM}$. We can also express the velocity as a magnitude and direction. $v_{CM} = \sqrt{v_{CM,x}^2 + v_{CM,y}^2} = 24.6$ km/s. To calculate the angle, note that since both $v_{CM,x}$ and $v_{CM,y}$ are negative, the angle is in the third quadrant. This means we must add 180° to the result a calculator gives to get the correct angle. $\theta_{VCM} = \tan^{-1} \dfrac{v_{CM,y}}{v_{CM,x}} = 225°.$

10-69. $\mathbf{v}_{CM} = \dfrac{\sum_i m_i \mathbf{v}_i}{\sum m_i} = \dfrac{(1500\text{ kg})(25\text{ m/s} - 15\text{ m/s})}{3000\text{ kg}}\mathbf{i} = 5.0\mathbf{i}\text{ m/s.}$

The kinetic energy carried by the CM is $K_{CM} = \dfrac{\left(\sum m_i\right)v_{CM}^2}{2} = \dfrac{(3000\text{ kg})(5.0\text{ m/s})^2}{2} = 3.8 \times 10^4\,\text{J.}$

The total kinetic energy is $K_{TOT} = \dfrac{\sum_i m_i v_i^2}{2} = \dfrac{(1500\text{ kg})[(25\text{ m/s})^2 + (15\text{ m/s})^2]}{2} = 6.38 \times 10^5\,\text{J.}$

10-73. For helium, $m_{He} = 4.00\text{ u} \times 1.66 \times 10^{-27}\text{ kg/u} = 6.64 \times 10^{-27}\text{kg.}$

$K_{TOT} = N_A \dfrac{m_{He} v^2}{2} = \dfrac{(6.02 \times 10^{23})(6.64 \times 10^{-27}\text{ kg})(1.4 \times 10^3\text{m/s})^2}{2} = 3.9 \times 10^3\,\text{J.}$

For oxygen molecules, $m_{O_2} = 2 \times 16.0\text{ u} \times 1.66 \times 10^{-27}\text{ kg/u} = 5.31 \times 10^{-26}\text{kg.}$

$K_{TOT} = N_A \dfrac{m_{O_2} v^2}{2} = \dfrac{(6.02 \times 10^{23})(5.31 \times 10^{-26}\text{ kg})(500\text{ m/s})^2}{2} = 4.0 \times 10^3\,\text{J.}$ Note that these energies are basically identical.

10-75. Assume $\mathbf{v}_{CM} = 0$. Since $\mathbf{v}_{CM} = \dfrac{\sum m_i \mathbf{v}_i}{\sum m_i}$, this means $M_J \mathbf{v}_J = - M_S \mathbf{v}_S.$ In terms of speeds,

$v_S = \dfrac{M_J v_J}{M_S}.$ The ratio of kinetic energies is $\dfrac{K_S}{K_J} = \dfrac{M_S v_S^2}{M_J v_J^2} = \dfrac{M_J}{M_S} = 9.54 \times 10^{-4}.$

10-77. Assume the hunter is at rest when he fires the first bullet. Then he gets a recoil speed of $v_1 = \dfrac{m v_B}{M}$, where m is the mass of the bullet, v_B is the speed of the bullet, and M is the mass of the hunter. When the second bullet is fired, it has a speed v_B relative to the already moving hunter and produces a recoil speed v_1 relative to the hunter, so now the hunter's speed relative to the ground is $2v_1$. Each bullet adds an increment v_1 to the hunter's speed relative to the ground, so after firing n bullets his speed is $v_n = n\dfrac{m v_B}{M}.$ Thus $v_n = 10 \times \dfrac{(0.015\text{ kg})(600\text{ m/s})}{80\text{ kg}} = 1.1\text{ m/s.}$

10-79. The magnitude of the net horizontal force on each child is 200 N, so the magnitudes of their horizontal accelerations are

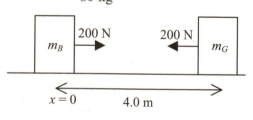

$a_B = \dfrac{F_{net}}{m_B} = \dfrac{200\text{ N}}{40\text{ kg}} = 5.0\text{ m/s}^2$

$a_G = \dfrac{F_{net}}{m_G} = \dfrac{200\text{ N}}{30\text{ kg}} = 6.7\text{ m/s}^2$

The accelerations are in opposite directions. Taking x to point right, $\mathbf{a}_B = 5.0\mathbf{i} \text{ m/s}^2$, $\mathbf{a}_G = -6.71 \text{ m/s}^2$. Since the net external force on the pair of children is zero, their mutual CM remains at rest. That is the point where they will meet. Taking $x = 0$ at the boy's initial position,

$$x_{CM} = \frac{\sum m_i x_i}{\sum m_i} = \frac{(30 \text{ kg})(4.0 \text{ m})}{70 \text{ kg}} = 1.7 \text{ m}. \text{ They will meet 1.7 m to the right of the boy.}$$

10-81.   The initial momentum is $\mathbf{p}_i = m_{car}\mathbf{v}_{car} + m_{truck}\mathbf{v}_{truck}$. The two vehicles are at rest after the collision, so $\mathbf{p}_f = 0$. By conservation of momentum, $\mathbf{p}_i = \mathbf{p}_f \Rightarrow m_{car}\mathbf{v}_{car} + m_{truck}\mathbf{v}_{truck} = 0$, which means

$\mathbf{v}_{truck} = -\dfrac{m_{car}\mathbf{v}_{car}}{m_{truck}}$. Since speed is the magnitude of velocity, $v_{truck} = \dfrac{m_{car}v_{car}}{m_{truck}}$, and we get

$$v_{truck} = \frac{m_{car}(40 \text{ km/h})}{5m_{car}} = 8.0 \text{ km/h}.$$

10-85.   To have a concrete starting point for the analysis, assume that the cat is initially at the center of the board so she's at the CM. (It doesn't really matter, but this assumption makes the calculations simpler.) Call this position $x = 0$. Since the net force on the cat and board is zero, the CM of the cat-board system must remain fixed as the cat walks. Suppose the cat has walked to a position x_C to the right of the initial position. Since the CM of the system must remain at $x = 0$, the CM of the board must have moved to a position $-x_B$ to the left of $x = 0$. (x_B is the *distance* to the position of the board's CM.) Mathematically, this is means $x_{CM} = 0 = \dfrac{m_C x_C - m_B x_B}{m_C + m_B}$. We're told that the cat walks 1.0 m along the board, so $x_C + x_B = 1.0$ m. (Remember that the board must move under the cat.) Thus $x_B = 1.0 \text{ m} - x_C$. Substituting into the equation for x_{CM} gives $\dfrac{m_C x_C - m_B(1 - x_C)}{m_C + m_B} = 0$

$\Rightarrow x_C(m_C + m_B) = (1.0)m_B \Rightarrow x_C = \dfrac{(1.0)m_B}{(m_C + m_B)}$. Substituting numbers gives

$$x_C = \frac{(1.0 \text{ m})(5.0 \text{ kg})}{8.5 \text{ kg}} = 0.59 \text{ m}. \text{ The cat moves } \underline{59 \text{ cm}} \text{ relative to the water.}$$

CHAPTER 11   COLLISIONS

11-5.   28 g = 2.8×10^{-2} kg   70 cm = 0.7 m

$p = mv = 0$ $p' = 2.8 \times 10^{-2}$ kg $\times$ 450 m/s = 12.6 kg-m/s

Impulse = Δp = 12.6 kg-m/s – 0 = 12.6 kg-m/s

$v'^2 = v^2 + 2ax$ so $a = \dfrac{v'^2 - v'^2}{2x} = \dfrac{(450\,\text{m/s})^2}{2 \times 0.7\text{m}} = 1.45 \times 10^5$ m/s^2

$t = \dfrac{v}{a} = \dfrac{450\,\text{m/s}}{1.45 \times 10^5\,\text{m/s}^2} = 3.10 \times 10^{-3}$ s = time to accelerate

$F = \dfrac{\Delta p}{\Delta t} = \dfrac{12.6\text{kg-m/s}}{0.0031\ \text{s}} = 4.06 \times 10^3$ N

(Compare this with $F = ma$: $F = 2.8 \times 10^{-2}$ kg $\times$ 1.45×10^5 m/s^2 = 4.06×10^3 N)

11-7. $F = \dfrac{\Delta p}{\Delta t}$ 50 km/hr = 13.9 m/s

$p = mv$ = 10 kg $\times$ 13.9 m/s = 139 kg-m/s $p' = 0$

Impulse = Δp = 0 – 139 kg-m/s = –139 kg-m/s

$|F| = \dfrac{139\text{kg-m/s}}{0.10\text{s}}$ = 1390 N, or about 310 lb. It is unlikely that the mother could hold on to the child.

11-9.   Because the collision is elastic the initial velocity of the ball will be 15 m/s and the final velocity will be –15 m/s. The mass of the ball is 60 g = 0.06 kg. The impulse is the change in momentum

$p = mv$ = 0.06 kg $\times$ 15 m/s = 0.9 kg-m/s p' = 0.06 kg $\times$ –15 m/s = –0.9 kg-m/s

Impulse = Δp = –0.9 kg-m/s – 0.9 kg-m/s = –1.8 kg-m/s

To stop the ball $a = \dfrac{v'^2 - v^2}{2x} = \dfrac{0 - (15\ \text{m/s})^2}{2 \times 0.005\ \text{m}} = -2.25 \times 10^4$ m/s^2

Time to stop the ball is $\dfrac{\Delta v}{a} = \dfrac{-15\,\text{m/s}}{-2.25 \times 10^4\,\text{m/s}^2} = t = 6.67 \times 10^{-6}$ s This is the same time needed to accelerate the ball.

$F = \dfrac{\Delta p}{\Delta t} = \dfrac{-1.8\ \text{kg-m/s}}{2 \times (6.67 \times 10^{-4}\text{s})} = -1.35 \times 10^3$ N

(Check with $F = ma$: $F = 0.06$ kg $\times$ –2.25 $\times 10^4$ m/s^2 = –1.35 $\times 10^3$ N)

11-11. $p = mv = 0$ p' = 0.45 kg $\times$ 18 m/s = 8.1 kg-m/s

Impulse to ball = $\Delta p = p' - p$ = 8.1 kg-m/s – 0 kg-m/s = 8.1 kg-m/s

Impulse is $F\Delta t = \Delta p$ so $\Delta t = \dfrac{\Delta p}{F} = \dfrac{8.1\ \text{kg-m/s}}{180\ \text{N}} = 0.045$ s

11-13.   Impulse $I = \displaystyle\int_0^{\Delta t} F dt = \int_0^{3s} (3.0t + 0.5t^2)\,dt = \int_0^{3s} 3.0t\,dt + \int_0^{3s} 0.5t^2\,dt = \left.\dfrac{3t^2}{2}\right|_0^{3s} + \left.\dfrac{0.5t^3}{3}\right|_0^{3s} = 18$ kg • m/s

11-15. $120 \text{ km/hr} = 33.3 \text{ m/s}$

Time interval of contact

$$\Delta t = \frac{2.5 \text{ m}}{\text{horizontal speed}} = \frac{2.5 \text{ m}}{33.3 \text{ m/s} \times \cos(3)} = \underline{7.52 \times 10^{-2} \text{ s}}$$

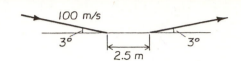

Magnitude of average force

$$F = \frac{\Delta p}{\Delta t} = \frac{2 \left[600 \text{ kg} \times 33.3 \text{ m/s} \times \sin(3) \right]}{7.52 \times 10^{-2} \text{ s}} = \underline{2.8 \times 10^{4} \text{ N}}$$

11-17. (a) Using equations 11.13 and 11.14 (pp. 345-346), velocity of the projectile after collision

$$v_1' = \frac{m_1 - m_2}{m_1 + m_2} v_1 = \frac{-0.06 \text{ kg}}{0.18} \times 0.8 \text{ m/s} = \underline{-0.27 \text{ m/s}}$$

velocity of target after collision

$$v_2' = \frac{2m_1}{m_1 + m_2} v_1 = \underline{0.53 \text{ m/s}}$$

(b) Initial kinetic energy of projectile:

$$K_1 = \frac{1}{2} m_1 v_1^2 = \frac{1}{2}(0.06 \text{ kg})(0.8 \text{ m/s})^2 = \underline{1.9 \times 10^{-2} \text{ J}}$$

Target: $K_2 = \underline{0 \text{ J}}$

Final kinetic energy of:

Projectile: $K_1' = \frac{m_1 (v_1')^2}{2} = \frac{1}{2}(0.06 \text{ kg})(0.27 \text{ m/s})^2 = \underline{2.2 \times 10^{-3} \text{ J}}$

Target: $K_2' = \frac{m_2 (v_2')^2}{2} = \frac{1}{2}(0.12 \text{ kg})(0.53 \text{ m/s})^2 = \underline{1.7 \times 10^{-2} \text{ J}}$

11-21. In an elastic collision kinetic energy and momentum are conserved and the velocity of the nail just after impact is given by eq. 11.14:

$$v_2' = \frac{2m_1}{m_1 + m_2} v_1 = \frac{2 \times 0.5 \text{ kg}}{0.012 \text{ kg} + 0.5 \text{ kg}} \times 5.0 \text{ m/s} = 9.77 \text{ m/s}$$

The kinetic energy of the nail $K_N = \frac{1}{2} m_2 v_2'^2 = \frac{1}{2}(0.012$

$\text{kg})(9.77 \text{ m/s})^2 = \underline{0.57 \text{ J}}$

11-23. The first ball hits a stationary block, but after that first collision the block is moving and Equations 11.13 and 11.14 are not applicable. We must instead solve the problem of a mass m_1 moving with initial velocity v_1 striking a mass m_2 moving with velocity v_2. After the collision the masses have velocities v'_1 and v'_2 respectively. Then conservation of momentum and kinetic energy require

$$m_1 v_1 + m_2 v_2 = m_1 v_1' + m_2 v_2'$$

$$\frac{m_1 v_1^2}{2} + \frac{m_2 v_2^2}{2} = \frac{m_1 (v_1')^2}{2} + \frac{m_2 (v_2')^2}{2}$$

The solution can be simplified by describing the motion in a frame of reference moving at velocity v_2. In this frame, m_2 is at rest and Equations 11.13 and 11.14 can be used. Define new velocities:

$$V_1 = v_1 - v_2$$
$$V_1' = v_1' - v_2$$
$$V_2' = v_2' - v_2$$

In this frame $V_2 = 0$ and the final velocities are

$$V_1' = \frac{m_1 - m_2}{m_1 + m_2} V_1$$

$$V_2' = \frac{2m_1}{m_1 + m_2} V_2$$

To get the final velocities in the original frame, substitute the velocity definitions:

$$v_1' - v_2 = \frac{m_1 - m_2}{m_1 + m_2}(v_1 - v_2) \Rightarrow v_1' = \frac{(m_1 - m_2)v_1 + 2m_2 v_2}{m_1 + m_2}$$

$$v_2' - v_2 = \frac{2m_1}{m_1 + m_2}(v_1 - v_2) \Rightarrow v_2' = \frac{2m_1 v_1 + (m_2 - m_1)v_2}{m_1 + m_2}$$

(If $v_2 = 0$, these reduce to Equations 11.13 and 11.14.) In this problem, $v_1 = v$ and $m_2 = 40m_1$ so the sum of the masses is $41m_1$ and the magnitude of the difference is $39m_1$. For the first collision, $v_2 = 0$ and

$$v_{2,1}' = \frac{2m_1 v}{41m_1} = \frac{2}{41}v = 0.0488v$$

For the next collision, this becomes the initial velocity of m_2 and the final velocity is

$$v_{2,2}' = \frac{2m_1 v + 39m_1(0.0488v)}{41m_1} = 0.0952v$$

Continuing the process gives

$$v_{2,3}' = \frac{2m_1 v + 39m_1(0.0952v)}{41m_1} = 0.139v$$

$$v_{2,4}' = \frac{2m_1 v + 39m_1(0.139v)}{41m_1} = 0.181v$$

$$v_{2,5}' = \frac{2m_1 v + 39m_1(0.181v)}{41m_1} = \underline{0.222v}$$

11-27.   The collision is elastic. In a reference frame moving at 55 km/hr the front car is at rest and we can use

$$v_1' = \frac{m_1 - m_2}{m_1 + m_2} v_1 \qquad v_2' = \frac{2m_1}{m_1 + m_2} v_1$$

60 km/hr = 16.7 m/s   55 km/hr = 15.3 m/s This give us

$v_1 = 1.4$ m/s $v_2 = 0$ $m_1 = 1200$ kg $m_2 = 1000$ kg

$$v_1' = \frac{1200 \text{ kg} - 1000 \text{ kg}}{1200 \text{ kg} + 1000 \text{ kg}} \times 1.4 \text{ m/s} = 0.13 \text{ m/s and } v_2' = \frac{2(1200 \text{ kg})}{1200 \text{ kg} + 1000 \text{ kg}} \times 1.4 \text{ m/s} = 1.5 \text{ m/s}$$

Changing back to the original frame of reference (adding 15.3 m/s to each velocity)

$$\underline{v_1' = 15 \text{ m/s}} \text{ and } \underline{v_2' = 17 \text{ m/s}}$$

11-29. This is an elastic collision and we can use equations 11.13 and 11.14

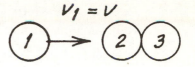

$$v_1' = \frac{m_1 - m_2}{m_1 + m_2} v_1 \quad v_2' = \frac{2m_1}{m_1 + m_2} v_1$$

Collision between first and second balls:

Equal masses gives $v_1' = \frac{m_1 - m_2}{m_1 + m_2} v_1 = 0$ and $v_2' = \frac{2m_1}{m_1 + m_2} v_1 = v_1$

Collision between second and third balls:

$v_2' = \frac{m_2 - m_3}{m_2 + m_3} v_2 = 0$ and $v_3' = \frac{2m_2}{m_2 + m_3} v_2 = v_1$

Last ball moves with velocity $v_1 = v$, other two are stationary.

11-35. The 1:10 slope implies a right triangle with an angle of $\tan^{-1}(1/10) = 5.71°$. With the hypotenuse as the ramp, moving 12 m down the hypotenuse means a drop $h = 12 \times \sin(5.71°) = 1.19$ m. In dropping 1.2 m the automobile gains kinetic energy of $mgh = \frac{1}{2}mv^2$. At the bottom the first automobile has a speed $v_1 = \sqrt{2gh} = \sqrt{2 \times 9.81 \text{ m/s}^2 \times 1.19 \text{ m}} = 4.83$ m/s. For the second automobile $v_2 = 0$.

After the collision eqs. 11.13 and 11.14 give

$$v_1' = \frac{m_1 - m_2}{m_1 + m_2} v_1 = \frac{1400 \text{ kg} - 800 \text{ kg}}{1400 \text{ kg} + 800 \text{ kg}} \times 4.83 \text{ m/s} = 1.32 \text{ m/s} \text{ and}$$

$$v_2' = \frac{2m_1}{m_1 + m_2} v_1 = \frac{2 \times 1400 \text{ kg}}{1400 \text{ kg} + 800 \text{ kg}} \times 4.83 \text{ m/s} = 6.15 \text{ m/s}$$

Frictional force for automobile 2: $f = \mu N = 0.9 \times 800 \text{ kg} \times 9.81 \text{ m/s}^2 \times \cos(5.71°) = 7828$ N
Gravitational force on automobile 2: $F_g = 800 \text{ kg} \times 9.81 \text{ m/s}^2 \times \sin(5.71°) = 781$ N
Net force on automobile 2: $F_2 = f + F_g = -7828 \text{ N} + 781 \text{ N} = -7047$ N, where the positive direction is taken to point down the slope. Acceleration of automobile 2: $a_2 = -7047 \text{ N}/800 \text{ kg} = -8.81 \text{ m/s}^2$
Acceleration of automobile 1: $a_1 = g\sin(5.71°) = 9.81 \text{ m/s}^2 \times \sin(5.71°) = 0.976 \text{ m/s}^2$

The cars will collide when they have traveled the same distance $x = vt + \frac{at^2}{2}$

$$\Rightarrow v_1 t + \frac{a_1 t^2}{2} = v_2 t + \frac{a_2 t^2}{2} \Rightarrow 1.32 + \frac{0.976t}{2} = 6.15 - \frac{8.81t}{2}$$

$$t = \frac{9.66}{9.79} \text{ s} = 0.987 \text{ s}$$

Automobiles travel $x = (1.32 \text{ m/s})(0.987 \text{ s}) + \frac{(.976)(0.987)^2}{2} = 1.78$ m

11-37. Suppose the fist and the block begin moving together at the same speed after impact so the collision is completely inelastic. The final

velocity is $v' = \dfrac{m_1 v_1}{m + m_2}$, where m_1 = mass of fist

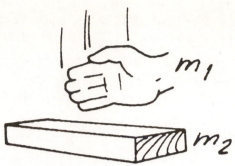

and m_2 = mass of block. The energy available to break the block is the difference between the initial kinetic energy and the final kinetic energy:

$$E = K - K' = \frac{m_1 v_1^2}{2} - \frac{(m_1 + m_2)v'^2}{2} = \frac{m_1 v_1^2}{2} - \frac{(m_1 + m_2)}{2}\left(\frac{m_1 v_1}{m_1 + m_2}\right)^2 = \frac{m_1 m_2 v_1^2}{2(m_1 + m_2)}$$

Since the block is considerably heavier than the fist, i.e., $m_2 \gg m_1$, this becomes

$E \approx \dfrac{m_1 v_1^2}{2} = \dfrac{0.4 \times 12^2}{2}$ J = $\underline{29\ J}$, which is enough to break the block.

11-41. Conservation of momentum for a totally inelastic collision $m_1 v_1 + m_2 v_2 = (m_1 + m_2)v'$
For the woman, the first throw just causes her to recoil:
$0 = m_1 v_1' + m_2 v_2' = 5\ \text{kg} \times 2.5\ \text{m/s} + 75\ \text{kg} \times v_w'$ $\underline{v_w' = -0.167\ \text{m/s}}$
For the man, first throw
$0 + 5\ \text{kg} \times 2.5\ \text{m/s} = (65\ \text{kg} + 5\ \text{kg}) \times v_m'$ $\underline{v_m' = 0.179\ \text{m/s}}$
For the man, second throw. The ball's velocity relative to the man is –3 m/s, so its velocity relative to the ice is $(-3 + 0.179)$ m/s $= -2.82$ m/s. Then
$70\ \text{kg} \times 0.179\ \text{m/s} = 5\ \text{kg} \times -2.82\ \text{m/s} + 65\ \text{kg} \times v_m'$ $\underline{v_m' = 0.410\ \text{m/s}}$
For the woman, second throw
$75\ \text{kg} \times (-0.167\ \text{m/s}) + 5\ \text{kg} \times -2.82\ \text{m/s} = 80\ \text{kg} \times v_w'$ $\underline{v_w' = -0.333\ \text{m/s}}$

11-45. (a) 80 km/h = 22.2 m/s. By conservation of momentum,

$$v' = \frac{m_1 v_1 - m_2 v_2}{m_1 + m_2} = \frac{1400\ \text{kg} - 540\ \text{kg}}{1400\ \text{kg} + 540\ \text{kg}} \times 22.2\ \text{m/s} = \underline{9.84\ \text{m/s}}$$

(b) Before collision:

$$K = \frac{1}{2}m_1 v_1^2 + \frac{1}{2}m_2 v_2^2 = \frac{1}{2}(1400\ \text{kg} + 540\ \text{kg})(22.2\ \text{m/s})^2 = \underline{4.78 \times 10^5\ \text{J}}$$

After collision:

$$K' = \frac{1}{2}(m_1 + m_2)(9.84\ \text{m/s})^2 = \underline{9.4 \times 10^4\ \text{J}}$$

(c) To determine the accelerations relative to the ground, it is necessary to know how far each car moves during the collision. This is difficult to determine in the frame relative to the ground but it is easy to do in a reference frame moving with the center of mass. In the CM frame, the distance each car travels during the collision is the crumple distance of 0.60 m. The center of mass moves with constant velocity, so the accelerations in the center of mass frame will be the same as the accelerations relative to the ground. The final velocity of each car in the CM frame is zero, and the initial velocities are found by subtracting the velocity of the CM relative to the ground from the initial velocities of the cars. The velocity of the CM is the same as the final velocity of the combined mass after the collision which was found in (a). Thus $v_{CM} = 9.84$ m/s in the direction of motion of the 1400 kg car. Then the initial velocities in the CM frame are $v_1 = (22.2 - 9.84)$ m/s = 12.4 m/s for the 1400 kg car and and $v_2 = (-22.2 - 9.84)$ m/s $= -32.0$ m/s for the 540 kg car. The magnitudes of the accelerations are

$$|a_1| = \left|\frac{v_f^2 - v_1^2}{2x}\right| = \left|\frac{0 - 12.4^2}{2(0.60)}\right| = 128 \text{ m/s}^2$$

$$|a_2| = \left|\frac{v_f^2 - v_2^2}{2x}\right| = \left|\frac{0 - 32.0^2}{2(0.60)}\right| = 853 \text{ m/s}^2$$

11-47. The "energy available for inelastic reactions" is that part of the total kinetic energy that is not associated with the motion of the center of mass. The kinetic energy of the CM is $K_{CM} = \dfrac{Mv_{CM}^2}{2}$, where M is the total mass of the particles in the system. By equation 11.27, $v_{CM} = \dfrac{v}{2}$ for two identical particles when one is at rest and the other is moving with speed v. Since one proton is moving, the total kinetic energy is $K = \dfrac{m_p v^2}{2}$. The kinetic energy of the CM is

$$K_{CM} = \frac{(2m_p)(v/2)^2}{2} = \frac{m_p v^2}{4} = \frac{K}{2}. \text{ Thus the available energy is}$$

$$E = \frac{K}{2} = \frac{8.0 \times 10^{-13}\text{ J}}{2} = 4.0 \times 10^{-13}\text{ J}$$

11-49. From eq. 11.27 the velocity of the center of mass is $v_{CM} = \dfrac{m_1 v_1 + m_2 v_2}{m_1 + m_2}$ and $v_{CM}' = v_{CM}$

$$v_{CMx} = \frac{m_1 v_1}{m_1 + m_2} = 16 \text{ m/s}, v_{CMy} = \frac{m_1 v_2}{m_1 + m_2} = 8.1 \text{ m/s} \Rightarrow v_{CM} = \sqrt{16^2 + 8.1^2} = 17.9 \text{ m/s}$$

(a) In the CM frame of reference:
Before:

$$K_{CM} = \frac{1}{2}(m_1 + m_2)v_{CM}^2 = \frac{(2400 \text{ kg})(17.9 \text{ m/s})^2}{2} = 3.9 \times 10^5 \text{ J}$$

Since the CM travels at constant velocity, K_{CM} is still 3.9×10^5 J after the collision.
(b) Before the collision:

$$K = \frac{1}{2}m_1 v_1^2 + \frac{1}{2}m_2 v_2^2 = \frac{(1100 \text{ kg})(34 \text{ m/s})^2}{2} + \frac{(1300 \text{ kg})(15 \text{ m/s})^2}{2} = 7.8 \times 10^5 \text{ J}$$

After the collision, the vehicles are moving at v_{CM}, so their total kinetic energy is the same as K_{CM} found in (a): $K' = 3.9 \times 10^5$ J.

11-53. (a) Treat this as an inelastic collision where the two bodies do not stay together.
Conservation of momentum $\qquad m_1 v_1 + m_2 v_2 = m_1 v_1' + m_2 v_2'$

$$(0.015 \text{ kg})(600 \text{ m/s}) = (0.015 \text{ kg})v_1' + (0.3 \text{ kg})(8 \text{ m/s}) \quad v_1' = \frac{9 \text{ kg}\cdot\text{m/s} - 2.4 \text{ kg}\cdot\text{m/s}}{0.015 \text{ kg}} = 440 \text{ m/s}$$

$\Delta v_1 = 440 \text{ m/s} - 600 \text{ m/s} = -160 \text{ m/s}$
(b) Bullet:

$$K_{before} = \frac{1}{2}m_1 v_1^2 = \frac{1}{2}(0.015 \text{ kg})(600 \text{ m/s})^2 = 2700 \text{ J} \quad K_{after} = \frac{1}{2}m_1 v_1'^2 = \frac{1}{2}(0.015 \text{ kg})(440 \text{ m/s})^2$$

$= 1452 \text{ J}$
$\Delta K = K_2 - K_1 = 1452 \text{ J} - 2700 \text{ J} = -1248 \text{ J}$
(c) Wood:

$K_{before} = 0$ $K_{after} = \dfrac{1}{2}m_2v_2'^2 = \dfrac{1}{2}(0.3\ kg)(8\ m/s)^2 = 9.6\ J.$

$\Delta K = K_2 - K_1 = 9.6\ J - 0\ J = \underline{9.6\ J}$

(d) The missing energy (about 1240 J) is energy that shows up as heat in the bullet and block, work done in compression/deformation, and as noise.

11-55.   Once we know the speed of block and bullet after the collision we can use conservation of momentum to find the original speed of the bullet.

The block travels horizontally 1.4 m during the time it takes to fall 1.8 m

Time to fall: $t = \sqrt{\dfrac{2h}{g}} = \sqrt{\dfrac{2 \times 1.8\ m}{9.81\ m/s^2}} = 0.606\ s$

$x = vt$ so $v = \dfrac{x}{t} = \dfrac{1.4\ m}{0.606\ s} = 2.31\ m/s$

$(m_1 + m_2)v' = m_1v_1 + m_2v_2$ $(0.015\ kg + 4\ kg)(2.31\ m/s) = (0.015\ kg)v + 0$ $v = \underline{619\ m/s}$

11-57.   Momentum is conserved in the x and y directions. Since the masses and speeds of the hydrogen atoms are both the same

$mv_x = 2mv_x'$ and $mv_y = 2mv_y'$ so $v_x' = v_x/2$ and $v_y' = v_y/2$. Since $v_x = v_y = v$ the final speed will be

$v' = \sqrt{\left(\dfrac{v}{2}\right)^2 + \left(\dfrac{v}{2}\right)^2} = \dfrac{\sqrt{2}}{2}v$

The kinetic energy before and after the collision is

$\dfrac{1}{2}mv^2 + \dfrac{1}{2}mv^2 = \dfrac{1}{2}(2m)\left(\dfrac{\sqrt{2}}{2}v\right)^2$ Since 6.1×10^{-22} J is transferred to internal energy

$mv^2 - \dfrac{1}{2}m(v^2/2) = 6.1 \times 10^{-22}$ J or $\dfrac{1}{2}(1.66 \times 10^{-27}\ kg)v^2 = 6.1 \times 10^{-22}\ J \Rightarrow v = \underline{860\ m/s}$

11-59.   (a) Let one player have velocity
$\mathbf{v}_1 = 7.0\cos 65°\mathbf{i} - 7.0\sin 65°\mathbf{j}$ and other
$\mathbf{v}_2 = 7.0\cos 65°\mathbf{i} + 7.0\sin 65°\mathbf{j}$. From the
problem, the velocities make an angle
130° with respect to each other. Then, by
conservation of momentum
$m_1\mathbf{v}_1 + m_2\mathbf{v}_2 = (m_1 + m_2)\mathbf{v}'$

$\mathbf{v}' = \dfrac{m_1\mathbf{v}_1 + m_2\mathbf{v}_2}{m_1 + m_2}$

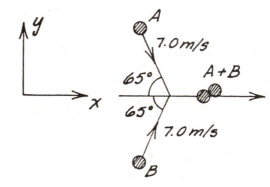

(here $m_1 = m_2 = 80$ kg.) Therefore,

$\mathbf{v}' = m(\mathbf{v}_1 + \mathbf{v}_2)/2m = \dfrac{1}{2}(\mathbf{v}_1 + \mathbf{v}_2)$

$= \dfrac{1}{2}(14.0\cos 65°\mathbf{i} + 0\mathbf{j}) = 7.0\cos 65°\mathbf{i} = \underline{3.0\mathbf{i}\ m/s}$

(b) $\Delta\mathbf{v} = \mathbf{v}_1 - \mathbf{v}'$ or $\mathbf{v}_2 - \mathbf{v}' = \pm 7.0\sin 65°\mathbf{j}$ m/s

The magnitude of the average acceleration $= |\Delta v/\Delta t|$

$\approx \left|\dfrac{7.0 \times \sin 65°}{0.080}\right| = \underline{79\ m/s^2}$

11-61. Let the x-axis point East, y-axis North. Let v_2 be the velocity of the other car. Then by the conservation of momentum $m_1\mathbf{v}_1 + m_2\mathbf{v}_2 = (m_1 + m_2)\mathbf{v}'$.

$\mathbf{v}_2 = v_2 \cos 40°\mathbf{i} - v_2 \sin 40°\mathbf{j}$.

Therefore,

$$\mathbf{v}' = \frac{(m_1\mathbf{v}_1 + m_2\mathbf{v}_2)}{m_1 + m_2}$$

$$= \left(\frac{900}{900+1200}\right)14\text{m/s } \mathbf{i} + \left(\frac{1200}{900+1200}\right)(v_2 \cos 40°\mathbf{i} - v_2 \sin 40°\mathbf{j})$$

$$= \frac{3}{7}(14\text{m/s})\mathbf{i} + \left(\frac{4}{7}\right)(v_2 \cos 40°\mathbf{i} - v_2 \sin 40°\mathbf{j})$$

Magnitude of $\mathbf{v}'$ is $\sqrt{\left[6 + \left(\frac{4}{7}\cos 40°\right)v_2\right]^2 + \left[\frac{4}{7}\sin 40°v_2\right]^2}$

$$= \left[36 + 5.25v_2 + \left\{\left(\frac{4}{7}\cos 40°\right)^2 + \left(\frac{4}{7}\sin 40°\right)^2\right\}v_2^2\right]^{1/2}$$

$$= [36 + 5.25v_2 + 0.327v_2^2]^{1/2}$$

The acceleration of the cars is $a = -\mu g$. From the skid marks, the magnitude of $\mathbf{v}'$ is

$$v' = \sqrt{-2a(x - x_0)} = \sqrt{2(\mu g)(x - x_0)}$$

$$= \sqrt{2(0.85 \times 9.81 \text{ m/s}^2)(17.4 \text{ m})} = 17.0 \text{ m/s}$$

By equating the two, we get

$$36 + 5.25v_2 + 0.327v_2^2 = (17.0)^2 = 290$$

$$\Rightarrow 0.327v_2^2 + 5.25v_2 - 254 = 0$$

This gives $v_2 = \left[-525 \pm \sqrt{(5.25)^2 + 4(0.327)(254)}\right]/(2 \times 0.327) = 21$ m/s (taking the positive root

only). So the speed of the other car is 21 m/s.

11-65. For particles of equal mass, conservation of momentum and kinetic energy give

$$m\mathbf{v} = m\mathbf{v}_1 + m\mathbf{v}_2 \Rightarrow \mathbf{v} = \mathbf{v}_1 + \mathbf{v}_2$$

$$\frac{mv^2}{2} = \frac{mv_1^2}{2} + \frac{mv_2^2}{2} \Rightarrow v^2 = v_1^2 + v_2^2$$

where $\mathbf{v}$ is the initial velocity (assuming the target is at rest) and $\mathbf{v}_1$ and $\mathbf{v}_2$ are the final velocities. Assume the angle between the final velocities is θ. Taking the dot product of the momentum equation gives

$$\mathbf{v} \cdot \mathbf{v} = v^2 = (\mathbf{v}_1 + \mathbf{v}_2) \cdot (\mathbf{v}_1 + \mathbf{v}_2) = v_1^2 + v_2^2 + 2\mathbf{v}_1 \cdot \mathbf{v}_2 = v_1^2 + v_2^2 + 2v_1v_2 \cos\theta$$

Combining this with the energy equation gives

$$v_1v_2 \cos\theta = 0 \Rightarrow \theta = 90°$$

The final velocities are perpendicular.

11-69.   Using equations 11.13 and 11.14, with $v_1' =$ velocity of (initially) moving car. $v_2' =$ velocity of stationary car,

$$v_1' = [(m_1 - m_2)/(m_1 + m_2)]v_1 = \left(\frac{1200 - 700}{1200 + 700}\right)10 \text{ km/h} = \underline{2.6 \text{ km/h}}$$

$$v_2' = [2m_1/(m_1 + m_2)]v_1 = [2(1200)/(1200 + 700)]10 \text{ km/h} = \underline{13 \text{ km/h}}$$

11-71.   Since kinetic energies are given, it will be convenient to work directly with those energies instead of velocities. Since the amount of kinetic energy lost is greatest in a totally inelastic collision, the magnitude of this energy loss is the total "energy available for inelastic processes". Since the protons are moving in opposite directions, conservation of momentum says

$$mv_1 - mv_2 = 2mv' \Rightarrow v' = \frac{v_1 - v_2}{2}$$

The final kinetic energy is

$$K' = \frac{1}{2}(2m)v'^2 = m\left(\frac{v_1 - v_2}{2}\right)^2 = \frac{m(v_1^2 - 2v_1v_2 + v_2^2)}{4}$$

In terms of the initial kinetic energies, $v_1^2 = \frac{2K_1}{m}, v_2^2 = \frac{2K_2}{m}$. Substituting gives

$$K' = \frac{K_1 + K_2}{2} - \sqrt{K_1K_2}$$

The change in kinetic energy is

$$\Delta K = K' - (K_1 + K_2) = -\left(\frac{K_1 + K_2}{2} + \sqrt{K_1K_2}\right)$$

Thus the total energy available for inelastic processes is

$$E = |\Delta K| = \frac{K_1 + K_2}{2} + \sqrt{K_1K_2}$$

$$\Rightarrow E = \frac{(8.0 + 4.0) \times 10^{-13}}{2} + \sqrt{8 \times 4} \times 10^{-13} \text{ J} = \underline{1.2 \times 10^{-13} \text{ J}}$$

11-73.   (a) For an elastic collision between two equal masses where the moving mass has $v_1 = 20$ m/s,

$$v_1' = \frac{m_1 - m_2}{m_1 + m_2}v_1 \quad v_2' = \frac{2m_1}{m_1 + m_2}v_1$$

$$v_1' = \frac{m - m}{m + m} \times 20 \text{ m/s} = \underline{0} \quad v_2' = \frac{2m}{m + m} \times 20 \text{ m/s} = \underline{20 \text{ m/s}}. \text{ The balls just exchange velocity.}$$

(b) Ball 1 lands at the base of the fence. Ball 2 takes $t = \sqrt{\frac{2h}{g}} = \sqrt{\frac{2 \times 1.5 \text{ m}}{9.81 \text{ m/s}^2}} = 0.553$ s to fall and

moves horizontally $x = vt = (20 \text{ m/s})(0.553 \text{ s}) = \underline{11 \text{ m from the base of the fence.}}$

11-75. 40 km/hr = 11.1 m/s. The velocity after the collision
$= v' = m_1 v_1/(m_1 + m_2)$
$= [(3.0 \times 10^4 t)(11.1)\text{m/s}]/(3.0 \times 10^4 + 8.0 + 10^5)t =$
0.40 m/s

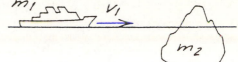

Therefore, ΔK/original K = fraction *converted* to inelastic energy
$$\left[\frac{1}{2}m_1 v_1^2 - \frac{1}{2}(m_1 + m_2)v'^2\right] \Big/ \left(\frac{1}{2}m_1 v_1^2\right) = 1 - \frac{(m_1 + m_2)v'^2}{m_1 v_1^2} = 1 - \frac{8.3 \times 10^5 (0.40)^2}{3.0 \times 10^4 (11.11)^2}$$
$= 1 - 0.036 = \underline{0.964}$

Fraction remaining as $K = \underline{0.036}$

11-77. (a) $v' = m_1 v/(m_1 + m_2) = (2 \times 10^9 \text{kg} \times 10^4 \text{m/s})/ (2 \times 10^9 + 5.98 \times 10^{24})\text{kg} = \underline{3.3 \times 10^{-12}}$ m/s
(b) The final recoil velocity is essentially zero, so all the initial kinetic energy is released for inelastic processes. This energy is
$$E = \frac{m_1 v_1^2}{2} = \frac{(2 \times 10^9 \text{ kg})(10^4 \text{ m/s})^2}{2} = 1 \times 10^{17} \text{ J} \times \frac{1 \text{ ton TNT}}{4.2 \times 10^9 \text{ J}} = \underline{2.4 \times 10^7 \text{ tons TNT (24 MT)}}$$

(c) Assume the meteorite came to rest over a distance equal to the depth of the crater, and assume that it was accelerated uniformly to a stop. Its average acceleration is $a = -\dfrac{v_1}{t}$. It comes

to rest in a distance $x = v_1 t + \dfrac{at^2}{2} = v_1 t - \dfrac{1}{2}\left(\dfrac{v_1}{t}\right)t^2 \Rightarrow t = \dfrac{2x}{v_1} = \dfrac{2(180 \text{ m})}{10^4 \text{m/s}} = 3.6 \times 10^{-2}$ s. The

magnitude of the impulse is $|\Delta p| = m_1 v_1 = 2 \times 10^9 \text{ kg} \times 10^4 \text{m/s} = 2 \times 10^{13} \text{ kg} \cdot \text{m/s}$.

$$F = \frac{\Delta p}{t} = \frac{2 \times 10^{13} \text{ kg} \cdot \text{m/s}}{0.036 \text{ s}} = \underline{5.6 \times 10^{14} \text{ N}}$$

11-81. (a) 70 km/h = 19.4 m/s, 100 km/h = 27.8 m/s. Momentum is conserved in both the N-S and E-W directions. Take N to be plus and E to be plus.
From equation 11.27 and $\mathbf{v_{CM}}' = \mathbf{v_{CM}}$ we have
$$v_{CME\text{-}W} = \frac{(1500 \text{ kg})(-27.8 \text{ m/s})\sin(30°)}{1500 \text{ kg} + 3500 \text{ kg}} = -4.17 \text{ m/s}$$

$$v_{CMN\text{-}S} = \frac{(3500 \text{ kg})(19.4 \text{ m/s}) + (1500 \text{ kg})(27.8 \text{ m/s})\cos(30°)}{1500 \text{ kg} + 3500 \text{ kg}} = 20.8 \text{ m/s}$$

$$v' = \sqrt{(20.8 \text{ m/s})^2 + (-4.17 \text{ m/s})^2} = \underline{21.2 \text{ m/s}}$$

$$\tan \theta = \frac{v_{CME\text{-}W}}{v_{CMN\text{-}S}} = \frac{-4.17 \text{ m/s}}{20.8 \text{ m/s}} \qquad \theta = -11.3°, \text{ or } \underline{11.3° \text{ West of North}}$$

(b) $K = \dfrac{m_1 v_1^2}{2} + \dfrac{m_2 v_2^2}{2} = \dfrac{3500 \times 19.4^2}{2} + \dfrac{1500 \times 27.8^2}{2} \text{ J} = 1.24 \times 10^6 \text{ J}$

$K' = \dfrac{(m_1 + m_2)v'^2}{2} = \dfrac{5000 \times 21.2^2}{2} \text{ J} = 1.12 \times 10^6 \text{ J}$

$\Delta K = K' - K = -1.2 \times 10^5 \text{ J}$, so the KE lost is $\underline{1.2 \times 10^5 \text{ J}}$.

12-5. (a) $\omega = 2\pi f = (2\pi)(33.33 \text{ rev/min})(1 \text{ min}/60 \text{ s}) = 3.49 \text{ radians/s}$. At the rim of the record, $v = \omega R = (0.15 \text{ m})(3.49 \text{ radians/s}) = \underline{0.52 \text{ m/s}}$. At the edge of the label, $R = 0.050 \text{ m}$, and $v = \underline{0.17 \text{ m/s}}$.

(b) $a_{centripetal} = \omega^2 R$. At the rim, $a_{centripetal} = (3.49 \text{ radians/s})^2 (0.15 \text{ m}) = \underline{1.8 \text{ m/s}^2}$. At the edge of the label, $R = 0.050 \text{ m}$, and $a_{centripetal} = \underline{0.61 \text{ m/s}^2}$.

12-7. $\omega = 2\pi f = (2\pi)(210 \text{ rev/min})(1 \text{ min}/60 \text{ s}) = \underline{22.0 \text{ radians/s}}$. $v = \omega R = (0.058 \text{ m})(22.0 \text{ radians/s}) = \underline{1.28 \text{ m/s}}$. At $R = 2.3 \text{ cm}$, $\omega = \dfrac{v}{R} = \dfrac{1.28 \text{ m/s}}{0.023 \text{ m}} = \underline{55.4 \text{ radians/s}}$.

$f = \dfrac{\omega}{2\pi} = \dfrac{55.7 \text{ radians/s}}{2\pi} = \underline{8.83 \text{ rev/s}}$.

12-9. $\omega = 2\pi f = (2\pi)(90 \text{ rev/min})(1 \text{ min}/60 \text{ s}) = 9.42 \text{ radians/s}$, or $\underline{9.4 \text{ radians/s}}$ to two significant figures. $v = \omega R = (0.10 \text{ m})(9.42 \text{ radians/s}) = \underline{0.94 \text{ m/s}}$.

$\overline{\alpha} = \dfrac{\Delta\omega}{\Delta t} = \dfrac{9.42 \text{ radians/s}}{5.0 \text{ s}} = \underline{1.9 \text{ radians/s}^2}$.

12-11. For aluminum, $v_{Al} = (100 \text{ m/min})(1 \text{ min}/60 \text{ s}) = 1.67 \text{ m/s}$. $v = 2\pi f R \Rightarrow f = \dfrac{v}{2\pi R}$

$= \dfrac{1.67 \text{ m/s}}{(2\pi)(3.0 \times 10^{-3}\text{m})} = \underline{88 \text{ rev/s}}$. For steel, $v_{st} = (20 \text{ m/min})(1 \text{ min}/60 \text{ s}) = 0.333 \text{ m/s}$.

$f = \dfrac{v}{2\pi R} = \dfrac{0.333 \text{ m/s}}{(2\pi)(2.5 \times 10^{-2}\text{m})} = \underline{2.1 \text{ rev/s}}$.

12-13. $\phi = C\left(t^2 - \dfrac{t^3}{4}\right)$. $\omega = \dfrac{d\phi}{dt} = C\left(2t - \dfrac{3t^2}{4}\right)$, $\alpha = \dfrac{d\omega}{dt} = C\left(2 - \dfrac{3t}{2}\right)$.

At $t = 0$, $\phi = 0$; $\omega = 0$; $\alpha = 40 \text{ radians/s}^2$.

At $t = 1.0 \text{ s}$, $\phi = (20 \text{ radians/s}^2)\left[(1.0 \text{ s})^2 - \dfrac{(1.0 \text{ s})^3}{4 \text{ s}}\right] = \underline{15 \text{ radians}}$;

$\omega = (20 \text{ radians/s}^2)\left[2(1.0 \text{ s}) - \dfrac{3(1.0 \text{ s})^2}{4 \text{ s}}\right] = \underline{25 \text{ radians/s}}$;

$\alpha = (20 \text{ radians/s}^2)\left[2.0 - \dfrac{3(1.0 \text{ s})}{2 \text{ s}}\right] = \underline{10 \text{ radians/s}^2}$.

At $t = 2.0 \text{ s}$, $\phi = (20 \text{ radians/s}^2)\left[(2.0 \text{ s})^2 - \dfrac{(2.0 \text{ s})^3}{4 \text{ s}}\right] = \underline{40 \text{ radians}}$;

$\omega = (20 \text{ radians/s}^2)\left[2(2.0 \text{ s}) - \dfrac{3(2.0 \text{ s})^2}{4 \text{ s}}\right] = \underline{20 \text{ radians/s}^2}$;

$\alpha = (20 \text{ radians/s}^2)\left[2.0 - \dfrac{3(2.0 \text{ s})}{2 \text{ s}}\right] = \underline{-20 \text{ radians/s}^2}$.

Note that the factor of 1/4 in the second term of ϕ must have units of s^{-1} for the expression to be dimensionally correct.

12-17. The final frequency $f = 7000$ rev/min $= 7000/60$ Hz $= 116.7$ Hz

$\Rightarrow$ final angular speed $\omega = 2\pi f = 2\pi$ rad $\times 116.7$ Hz $= 732.9$ rad/s

$\alpha = \dfrac{\omega - \omega_0}{t} = \dfrac{732.9 \text{ rad/s}}{1.2 \text{ s}} = \underline{611 \text{ rad/s}^2}$

The total angular displacement:

$\phi = \omega_0 t + \dfrac{1}{2}\alpha t^2 = \dfrac{1}{2} \times 611$ rad/s$^2 \times 1.2^2$ s$^2 = \underline{440 \text{ rad} = 70 \text{ revolutions}}$.

12-19. $\omega = (7200$ rev/min$)(2\pi$ radians/rev$)(1$ min/60 s$) = 754$ radians/s. $\bar{\alpha} = \dfrac{\Delta\omega}{\Delta t} = \dfrac{\omega - \omega_0}{\Delta t}$

$= \dfrac{754 \text{ radians/s-0}}{5.0 \text{ s}} = \underline{1.5 \times 10^2 \text{ radians/s}^2}$.

12-23. $\omega^2 - \omega_0^2 = 2\alpha\phi \Rightarrow \alpha = \dfrac{\omega^2 - \omega_0^2}{2\phi}$. $\omega_0 = 6.0$ rev/s $\times 2\pi$ radians/rev $= 12\pi$ radians/s.

$\phi = 5.0$ rev $\times 2\pi$ radians/rev $= 10\pi$ radians. $\alpha = \dfrac{0 - (12\pi \text{ radians/s})^2}{2(10\pi \text{ radians})} = -22.6$ radians/s^2, or

$\underline{-23 \text{ radians/s}^2}$ to two significant figures. $\alpha = \dfrac{\omega - \omega_0}{\Delta t} \Rightarrow \Delta t = \dfrac{\omega - \omega_0}{\alpha} = \dfrac{-12\pi \text{ radians/s}}{-22.6 \text{ radians/s}^2} = \underline{1.7 \text{ s}}$.

12-27. $\alpha = \dfrac{d\omega}{dt} \Rightarrow \int_{\omega_0}^{\omega} d\omega' = \int_0^t \alpha dt' \Rightarrow \omega = \omega_0 + \int_0^t C(t')^2 dt' = \omega_0 + \dfrac{Ct^3}{3}$. At $t = 3.0$ s,

$\omega = 8.0$ radians/s $+ \dfrac{(0.25 \text{ radians/s}^4)(3.0 \text{ s})^3}{3} = 10.25$ radians/s, or $\underline{\omega = 10 \text{ radians/s}}$ to two

significant figures. $\omega = \dfrac{d\phi}{dt} \Rightarrow \int_{\phi_0}^{\phi} d\phi' = \Delta\phi = \int_0^t \omega dt' \Rightarrow \Delta\phi = \int_0^t \left[\omega_0 + \dfrac{C(t')^3}{3}\right] dt' = \omega_0 t + \dfrac{Ct^4}{12}\Big|_0^{1.0}$

$\Delta\phi = \omega_0 t + \dfrac{Ct^4}{12} = (8.0 \text{ radians/s})(1.0 \text{ s}) + \dfrac{(0.25 \text{ radians/s}^4)(1.0 \text{ s})^4}{12} = \underline{8.0 \text{ radians}}$.

12-29. $\alpha = \dfrac{d\omega}{dt} = -A\omega \Rightarrow \dfrac{d\omega}{\omega} = -A dt$. Integrating both sides gives $\int_{\omega_0}^{\omega} \dfrac{d\omega'}{\omega'} = -A\int_0^t dt'$, from which

we get $\ln\omega'\Big|_{\omega_0}^{\omega} = -At \Rightarrow \ln\dfrac{\omega}{\omega_0} = -At$. Inverting the natural log gives $\underline{\omega = \omega_0 e^{-At}}$.

12-31. $I = MR^2 = (1.9 \times 10^6 \text{ kg})(38 \text{ m})^2 = \underline{2.74 \times 10^9 \text{ kg}\cdot\text{m}^2}$.

$\omega = 2\pi f = \dfrac{2\pi \text{ radians}}{\text{rev}} \times \dfrac{0.050 \text{ rev}}{\text{min}} \times \dfrac{1 \text{ min}}{60 \text{ s}} = 5.24 \times 10^{-3}$ radian/s, so

$K = \dfrac{I\omega^2}{2} = \dfrac{(2.74 \times 10^9 \text{ kg}\cdot\text{m}^2)(5.24 \times 10^{-3} \text{ radian/s})^2}{2} = \underline{3.8 \times 10^4 \text{ J}}$.

12-33. $I = \sum m_i R_i^2$ where R_i is the distance from the CM to the
oxygen atoms; $R_1 = R_2 = R$; $m_1 = m_2 = m$. So $I = 2mR^2$

$R = \sqrt{\dfrac{I}{2m}} = \sqrt{\dfrac{1.95 \times 10^{46} \text{ kg m}^2}{2 \times 2.66 \times 10^{-26} \text{ kg}}}$

$= 6.05 \times 10^{11}$ m. Thus, distance between atoms $= 2R$

$= \underline{1.21 \times 10^{10} \text{ m}}$.

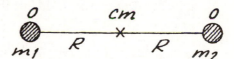

12-35. $I = \sum m_i R_i^2$, where R_i is the perpendicular distance to the axis. The only contribution is from two oxygen atoms off-axis. This gives:

$I = 2 \times 2.66 \times 10^{-26} \text{kg} \times (1.22 \times 10^{-10} \times \sin 65°)^2 \text{ m}^2 = \underline{6.50 \times 10^{-46} \text{ kg} \cdot \text{m}^2}$

12-37. $I = MR^2 = 4 \text{ kg} \times (0.33)^2 \text{ m}^2 = \underline{0.44 \text{ kg m}^2}$

12-41. The rotational frequency is 1 rev/day, so the angular speed is

$\omega = (1 \text{ rev/day})(2\pi \text{ radians/rev})(1 \text{day}/86{,}400 \text{ s}) = 7.27 \times 10^{-5}$ radian/s. The moment of inertia is

given, so $K = \dfrac{I\omega^2}{2} = \dfrac{(0.331)(5.98 \times 10^{24} \text{kg})(6.378 \times 10^6 \text{m})^2 (7.27 \times 10^{-5} \text{radians/s})^2}{2} =$

$\underline{2.13 \times 10^{29} \text{ J}}$.

12-47. The solution to this problem requires a result derived in Problem 53 and again in 57. The moment

of inertia of a thick spherical shell with inner radius R_1 and outer radius R_2 is $I = \dfrac{2M(R_2^5 - R_1^5)}{5(R_2^3 - R_1^3)}$.

The moment of inertia of the earth is $I_{earth} = I_{inner\ core} + I_{outer\ core}$. The inner core is a sphere, so

$I_{inner\ core} = \dfrac{2}{5}(0.22M_E)(0.54R_E)^2 = 0.0257 M_E R_E^2$. The outer core is a thick shell, so

$I_{outer\ core} = \dfrac{2(0.78M_E)\left[R_E^5 - (0.54R_E)^5\right]}{5\left[R_E^3 - (0.54R_E)^3\right]} = 0.353 M_E R_E^2$. Thus $\underline{I_{earth} = 0.379 M_E R_E^2}$.

12-49. $I = \int R^2 \, dm$.

Let the little mass element be ρdx
where ρ is the linear density of the
material. Then, $R = |x \sin \theta|$
(see diagram), hence,

$\int R^2 \, dM = \int_{x=-L/2}^{L/2} x^2 \sin^2 \theta \rho dx$

$= \rho \sin^2 \theta \int_{x=-L/2}^{L/2} x^2 \, dx$

$\rho \sin^2 \theta \left(\dfrac{x^3}{3}\right)_{-L/2}^{L/2} = \rho \dfrac{\sin^2\theta}{3}\left(\dfrac{L3}{8} + \dfrac{L3}{8}\right)$

$= \rho \dfrac{L}{3}\left(\dfrac{L^2}{4}\right) \sin^2 \theta = \underline{\dfrac{ML^2 \sin^2\theta}{12}}$

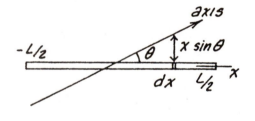

12-51. Take the y axis to be the axis of rotation. Imagine a strip of length
l and width dx a perpendicular distance x from the axis. The mass
of the strip is $dm = \rho l dx$, where ρ is the mass per unit area of the
plate. The moment of inertia of the strip is $dI = x^2 dm$. To find the
total moment of inertia, add up the dI's by integrating from $x = 0$
to $x = l$:

$I = \int_{x=0}^{x=l} dI = \int_{x=0}^{x=l} x^2 dm = \rho l \int_{x=0}^{x=l} x^2 dx = \dfrac{\rho l^4}{3}$. $\rho = M/l^2$, so

$I = \dfrac{Ml^2}{3}$.

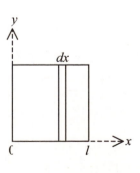

12-53. To solve this problem, we need the moment of inertia of a spherical shell with inner radius R_1 and outer radius R_2. Assume that the sphere is divided into thin shells of radius r and thickness dr. The volume of one of these shells is $dV = 4\pi r^2 dr$, so the mass contained in it is $dm = \rho dV = 4\pi \rho r^2 dr$, where ρ is the density. The moment of inertia of this thin shell is

$$dI = \frac{2}{3}(dm)r^2 = \frac{8\pi \rho r^4 dr}{3},$$ where we have used the moment of inertia of a thin spherical shell

from Table 12.3. To find the total moment of inertia, add up the dI's by integrating from $r = R_1$ to $r = R_2$:

$$I = \int_{r=R_1}^{r=R_2} dI = \frac{8\pi\rho}{3} \int_{R_1}^{R_2} r^4 dr = \frac{8\pi\rho}{15}\left(R_2^5 - R_1^5\right).$$

The volume of the shell can be found the same way: $V = \int_{r=R_1}^{r=R_2} dV = 4\pi \int_{R_1}^{R_2} r^2 dr = \frac{4\pi}{3}\left(R_2^3 - R_1^3\right)$,

so the density is $\rho = \dfrac{m}{V} = \dfrac{3m}{4\pi\left(R_2^3 - R_1^3\right)}$, and the moment of inertia of the thick shell is

$$I = \frac{2m\left(R_2^5 - R_1^5\right)}{5\left(R_2^3 - R_1^3\right)}.$$ Now we can proceed to find the moment of inertia of the peach.

$$I_{peach} = I_{pit} + I_{shell}.\ \ I_{pit} = \frac{2}{5}m_{pit}R_{pit}^2,\ \ I_{shell} = \frac{2m_{shell}\left(R^5 - R_{pit}^5\right)}{5\left(R^3 - R_{pit}^3\right)}.\ \ R_{pit} = 0.5R,\ m_{pit} = 0.05M,\ m_{shell} =$$

$$0.95M \Rightarrow I_{peach} = \frac{2}{5}(0.05M)\left(\frac{R}{2}\right)^2 + \frac{2(0.95M)\left[R^5 - \left(\dfrac{R}{2}\right)^5\right]}{5\left[R^3 - \left(\dfrac{R}{2}\right)^3\right]}.$$ Doing all the arithmetic finally

gives $\underline{I_{peach} = 0.005MR^2 + 0.421MR^2 = 0.426MR^2}$. (Because the mass of the pit is so small, we

need to keep three significant figures to see its effect on the total moment of inertia.)

12-57. (See also Problem 12-53.) $I = \int (r\sin\theta)^2 dm$.

$dm = \rho \cdot \sin\theta\ r^2 dr d\theta\, d\phi$, where

$$\rho = density = \frac{Mass\ of\ the\ shell}{Volume\ of\ the\ shell} = \frac{M}{\dfrac{4\pi}{3}(R_2^3 - R_1^3)} \Rightarrow$$

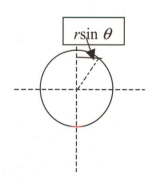

$$I = \rho \int (r\sin\theta)^2 \sin\theta\, r^2\ dr d\theta\ d\phi = \rho \int_0^{2\pi} d\phi \int_0^{\pi} \sin^3\theta\ d\theta \int_{R_1}^{R_2} r^4 dr =$$

$$\rho \cdot 2\pi \cdot \frac{4}{3} \cdot \frac{R_2^5 - R_1^5}{5}$$

$$\Rightarrow I = \frac{M}{\dfrac{4\pi}{3}(R_2^3 - R_1^3)} \cdot 2\pi \cdot \frac{4}{3} \cdot \frac{R_2^5 - R_1^5}{5} = \frac{2}{5}\frac{M(R_2^5 - R_1^5)}{R_2^3 - R_1^3}$$

12-59.   The volume element consisting of a disk of thickness dx has mass $dm = \rho dx(\pi r^2)$. But

$r = x \tan\theta$ where $\tan\theta = \dfrac{R}{l}$. Then

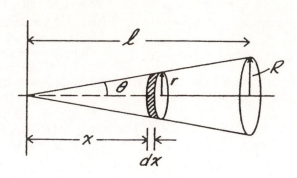

$$I = \int_0^l x^2 \; dm = \rho\pi\frac{R^2}{l^2}\int_0^l x^4 \; dx = \rho\pi\frac{R^2}{l^2}\left(\frac{x^5}{5}\right)_0^l$$

$$I = \rho\pi\frac{R^2}{l^2}\cdot\frac{l^5}{5} = \rho\pi\frac{R^2}{5}l^3.$$

The volume of a cone is $\dfrac{1}{3}\pi R^2 l$, so that

the mass is $\dfrac{1}{3}\rho\pi R^2 l$. Then $I = \left(\dfrac{1}{3}\rho\pi R^2 l\right)\dfrac{3}{5}l^2$

$$= \frac{3}{5}Ml^2$$

12-63.   Assume no slipping of the wheels.
Translational kinetic energy,

$$K_t = \frac{1}{2}Mv^2 = \frac{1}{2}1360\times\left(\frac{80}{3.6}\right)^2 = \underline{3.36\times10^5 \text{ J.}}$$

Angular speed of the wheel $\omega = \dfrac{v}{R} = \dfrac{80}{3.6\times0.381} = 58.3$ rad/s

Rotation kinetic energy, $E_R = 4\left(\dfrac{1}{2}I\omega^2\right) = 2I\omega^2$

$$E_R = 2\left[\frac{1}{2}(27.2)(0.381)^2\right](58.3)^2 = 1.34\times10^4 \text{ J}$$

Total kinetic energy $= \underline{3.49\times10^5 \text{ J.}}$
Rotational kinetic energy as percentage of the total energy

$$= \frac{E_R}{E_{\text{tot}}}\times100 = \frac{1.34\times10^4}{3.49\times10^5} = \underline{3.84\%}$$

12-65. I of sphere $= \dfrac{2}{5}MR^2$. $E = \dfrac{1}{2}I\omega^2$, so that $\dfrac{dE}{dt} = \dfrac{d}{dt}\left(\dfrac{1}{2}I\omega^2\right) = \dfrac{1}{2}I\dfrac{d}{dt}\omega^2$

$$\frac{dE}{dt} = \frac{1}{2}I\left[2\omega\left(\frac{d\omega}{dt}\right)\right] = I\omega\frac{d\omega}{dt} = \frac{2}{5}MR^2\omega\frac{d\omega}{dt}$$

$M = 1.5\times10^{30}$ kg, $R = 20{,}000$ m, $\omega = 2\pi(2.1)/\text{s} = 13.2/\text{s}$

$$\frac{d\omega}{dt} = 2\pi(1.0\times10^{-15}) \text{ rev/sec}^2 = -6.28\times10^{-15} \text{ rad/sec}^2$$

$|dE/dt| = $ rate energy decreasing

$$= \frac{2}{5}(1.5\times10^{30}\,\text{kg})(20{,}000^2\,\text{m}^2)(13.2/\text{s})(6.28\times10^{15}/\text{s}^2) = \underline{2.0\times10^{25} \text{ W}}$$

Pulsar stops when $K = 0$. $K = \dfrac{1}{2}I\omega^2 = \dfrac{1}{5}MR^2\omega^2$

$$= \frac{1}{5}(1.5\times10^{30}\,\text{kg})(20{,}000^2\,\text{m}^2)(13.2)^2/\text{s} = 2.1\times10^{40} \text{ J.}$$

If dE/dt = constant = $\Delta E/\Delta t$, then Δt = time taken for pulsar to stop
= $E/(\Delta E/\Delta t)$ = $(2.09 \times 10^{40}$ J$)/(1.99 \times 10^{25}$ W$)$ = 1.05×10^{15} s
$\Delta t = \underline{3.3 \times 10^7 \text{ yr}}$!

12-67.   Divide the plate into strips as shown. The y axis is the rotation
axis, and the perpendicular distance to the strip is x. The length of
the strip is $2\sqrt{R^2 - x^2}$, where R is the radius of the plate. The
thickness of the strip is dx. The mass of the strip
$dm = \rho(2\sqrt{R^2 - x^2})dx$, where ρ is the mass per unit area of the
plate. For each strip like the one shown for positive x, there is a
corresponding one for negative x. Thus the moment of inertia is
given by

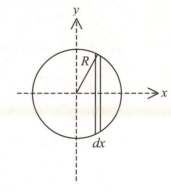

$$I = 2\int_0^R x^2 dm = 4\rho\int_0^R x^2\sqrt{R^2 - x^2}\,dx.$$

The integral can be found in standard tables. The result is

$$I = 4\rho\left[-\frac{x}{4}(R^2 - x^2)^{3/2} + \frac{R^2}{8}\left(x\sqrt{R^2 - x^2} + R^2\sin^{-1}\frac{x}{R}\right)\right]_0^R.$$

All terms in this are zero except for the inverse sine term
evaluated at $x = R$. The final result is

$I = \dfrac{\pi\rho R^4}{4}$. The mass per unit area is $\rho = \dfrac{M}{\pi R^2}$, so the final result

is $I = \underline{\dfrac{MR^2}{4}}$.

12-69.   Let z be the rotation axis. The mass element dm is a small strip of
length l and width dx located at a perpendicular distance x the
rotation axis. The mass of the strip is $dm = \rho l dx$, where ρ is the
mass per unit area of the plate. By the parallel axis theorem, the
moment of inertia of the strip for rotation about the z axis is

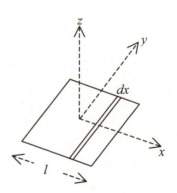

$dI = dI_{CM} + (dm)x^2 = \dfrac{l^2 dm}{12} + x^2 dm$. The strip shown is on the $+x$

side of the axis, and there is another corresponding strip on the
negative side. So when we integrate to add up the dI's, we can
integrate from $x = 0$ to $x = l/2$ and double the result.

$$I = 2\int_{x=0}^{x=l/2} dI = 2\int_{x=0}^{x=l/2} dm\left(\frac{l^2}{12} + x^2\right) = 2\rho l\int_{x=0}^{x=l/2}\left(\frac{l^2}{12} + x^2\right)dx$$

$$= 2\rho l\left[\frac{l^2 x}{12} + \frac{x^3}{3}\right]_0^{l/2} = 2\rho l\left(\frac{l^3}{24} + \frac{l^3}{24}\right)$$

$\rho = \dfrac{M}{l^2}$, so the final result is $I = \underline{\dfrac{Ml^2}{6}}$.

12-71. We could set this problem up exactly like 12-67, except that the negative half of the circular plate is missing. However, we can use proportional reasoning to get the result without actually repeating the calculations. Since the negative side of the plate is missing, we don't double the result of integrating from $x = 0$ to $x = R$, and that might lead us to expect that the moment of inertia is only half what was obtained in 12-67. However, the density is twice what it was in 12-67 because the area is only half as large, so the final result is that the moment of inertia of the semicircular plate is the same as that of the circular plate: $I = \dfrac{MR^2}{4}$. (The diagram in the text shows the y axis pointing along the direction labeled as x in 12-67, but this is irrelevant because the actual integration variable does not appear in the final result.)

12-77. The final tangential speed of a point on the rim of one of the tires is 80 km/h = 22.2 m/s. The final angular speed is $\omega = \dfrac{v}{r} = \dfrac{22.2 \text{ m/s}}{0.30 \text{ m}} = 74.1$ radians/s. The angular acceleration is

$\alpha = \dfrac{\omega - \omega_0}{t} = \dfrac{74.1 \text{ radians/s} - 0}{6.0 \text{ s}} = 12.3$ radians/s^2, or $\underline{12 \text{ radians/s}^2}$ to two significant figures. The

angle the tires rotated through is $\phi = \dfrac{\alpha t^2}{2} = \dfrac{(12.3 \text{ radians/s}^2)(6.0 \text{ s})^2}{2} = 221$ radians, which is

221 radians $\times \dfrac{1 \text{ rev}}{2\pi \text{ radians}} = \underline{35 \text{ revolutions}}$.

12-81. About the longitudinal axis through the center, $I_1 = 2 \times \dfrac{2}{5} MR^2 = \underline{\dfrac{4}{5} MR^2}$. To find the moment of

inertia about the transverse axis through the point of contact between the spheres, use the parallel axis theorem to find the moment of inertia for one sphere about an axis tangent to its surface, and

double that. $I_2 = 2(I_{CM} + Md^2)$, where $d = $ R. $I_2 = 2\left(\dfrac{2}{5} MR^2 + MR^2\right) = \underline{\dfrac{14}{5} MR^2}$.

12-85. When the cylinder is at the top of the ramp, its energy is $E = K + U = Mgh$, since it's at rest. At the bottom of the ramp, $E = K_{trans} + K_{rot}$, since its potential energy is now zero.

$K_{trans} + K_{rot} = \dfrac{Mv^2}{2} + \dfrac{I\omega^2}{2} = \dfrac{Mv^2}{2} + \dfrac{1}{2}\left(\dfrac{MR^2}{2}\right)\left(\dfrac{v}{R}\right)^2 = \dfrac{3Mv^2}{4}$. Using conservation of energy,

$\dfrac{3Mv^2}{4} = Mgh \Rightarrow v = \sqrt{\dfrac{4gh}{3}}$. For $h = 1.5$ m, $v = \sqrt{\dfrac{4(9.81 \text{ m/s}^2)(1.5 \text{ m})}{3}} = \underline{4.4 \text{ m/s}}$.

CHAPTER 13   DYNAMICS OF A RIGID BODY

13-1.   The crane boom of the crane makes an angle ϕ with respect to the horizontal direction. In the calculation of the torque about the pivot point P, which is given by $\tau = FR \sin \theta$, the angle is between the force **F** and the radial line **R**. The downward force **F** is due to the weight $m\mathbf{g}$ of the load. Since the load is directed vertically, a right triangle is formed between the boom, the rope supporting the load, and the horizontal direction. Thus, $\theta = 90° - \phi$.

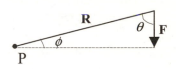

For $\phi = 20°$,

$$\tau_{max} = FR \sin \theta = mgR \sin (90° - 20°) = (500 \text{ kg})(9.81 \text{ m/s}^2)R \sin 70° = \underline{(4610 \text{ N} \bullet \text{m})R}$$

Using this result and assuming the maximum safe load is determined by this maximum safe torque, at $\phi = 40°$, the maximum safe load is

$$F_{Load} = \frac{\tau_{max}}{R \sin \theta} = \frac{(4610 \text{ N} \bullet \text{m})R}{R \sin(90° - \phi)} = \frac{4610 \text{ N} \bullet \text{m}}{\sin(90° - 40°)} = 6018 \text{ N}$$

The mass of this load is

$$m = \frac{F_{Load}}{g} = \frac{6018 \text{ N}}{9.81 \text{ m/s}^2} = \underline{613 \text{ kg}}$$

For $\phi = 60°$,

$$F_{Load} = \frac{\tau_{max}}{R \sin \theta} = \frac{(4610 \text{ N} \bullet \text{m})R}{R \sin(90° - \phi)} = \frac{4610 \text{ N} \bullet \text{m}}{\sin(90° - 60°)} = 9220 \text{ N}$$

and $m = \dfrac{F_{Load}}{g} = \dfrac{9220 \text{ N}}{9.81 \text{ m/s}^2} = \underline{940 \text{ kg}}$

13-3.   The pivot point for the wrench is at the bolt itself, so the radial line is along the length L of the handle of the wrench and the force will be applied perpendicular to the end of the wrench.
$\tau = FR \sin \theta = FL \sin 90° = FL$

$$F = \frac{\tau}{L} = \frac{62 \text{ N} \bullet \text{m}}{0.20 \text{ m}} = \underline{310 \text{ N}}$$

13-7.   The power delivered by the torque of the engine is $P = \tau \omega$, where ω is the angular speed in rad/s of the engine. When operating at a maximum torque τ_{max} of 203 N $\bullet$ m, the engine delivers a power of

$$P = \tau \omega = (203 \text{ N} \bullet \text{m})\left(4600 \frac{\text{rev}}{\text{min}} \right)\left(\frac{1 \text{ min}}{60 \text{ s}} \right)\left(\frac{2\pi \text{ rad}}{1 \text{ rev}} \right) = 9.8 \times 10^4 \text{ W} \left(\frac{1 \text{ hp}}{745.7 \text{ W}} \right) = \underline{130 \text{ hp}}$$

When running at a maximum power of 142 hp, the torque delivered by the engine is

$$\tau = \frac{P}{\omega} = \frac{142 \text{ hp}\left(\dfrac{745.7 \text{ J/s}}{1 \text{ hp}} \right)}{5750 \dfrac{\text{rev}}{\text{min}}\left(\dfrac{1 \text{ min}}{60 \text{ s}} \right)\left(\dfrac{2\pi \text{ rad}}{1 \text{ rev}} \right)} = \underline{176 \text{ N} \bullet \text{m}}$$

13-11. In Example 2, the angular speed was found for the meter stick by applying the principle of the conservation of mechanical energy. Every point on the stick, including the top end, has that angular speed as it rotates about the end initially in contact with the floor. The relationship between the tangential speed and the angular speed is $v = \omega r$. Therefore, the tangential speed at the end of the stick is

$$v = r\omega = (1.0 \text{ m})\sqrt{\frac{3g}{l}} = (1.0 \text{ m})\sqrt{\frac{3(9.81 \text{ m/s}^2)}{1.0 \text{ m}}} = \underline{5.4 \text{ m/s}}$$

If the stick is a 2 meter stick, then $v = r\omega = (2.0 \text{ m})\sqrt{\frac{3(9.81 \text{ m/s}^2)}{2.0 \text{ m}}} = \underline{7.7 \text{ m/s}}$

13-13. The power delivered (in watts) by the motor is the product of the motor torque (in N • m) and the angular speed (in rad/s). Since the angular speed is given in rev/min, a unit conversion is necessary.

$$P = \tau\omega = (0.65 \text{ N} \bullet \text{m})(3450 \text{ rev/min})\left(\frac{1 \text{ min}}{60 \text{ s}}\right)\left(\frac{2\pi \text{ rad}}{1 \text{ rev}}\right) = \underline{230 \text{ W}}$$

13-15. The work done in rotating the grinding table is equal to the product of the torque and the angle through which the table turns, $W = \tau\theta$. The angle is measured in radians, so a unit conversion is necessary.

$$W = \tau\theta = (250 \text{ N} \bullet \text{m})(1200 \text{ rev})\left(\frac{2\pi \text{ rad}}{1 \text{ rev}}\right) = \underline{1.9 \times 10^6 \text{ J}}$$

13-17. The work done by the motor is equal to the product of the torque and the angle through which the fan blades turn. Since the fan is starting from rest, the fan is undergoing an angular acceleration, so the angle has the time-dependence given. Therefore, to find the total work done, an integration over the elapsed time must be done.

For $t = 0$ to 1.0 s,

$$W = \tau\int_0^{1.0 \text{ s}}\frac{d\phi(t)}{dt}\,dt = \tau\int_0^{1.0 \text{ s}}\frac{d}{dt}(Ct^2)\,dt = 2C\tau\int_0^{1.0 \text{ s}}t\,dt = C\tau t^2\Big|_0^{1.0 \text{ s}}$$

$$= (7.5 \text{ rad/s}^2)(2.5 \text{ N} \bullet \text{m})(1.0 \text{ s})^2 = \underline{19 \text{ J}}$$

For $t = 0$ to 2.0 s,

$$W = C\tau t^2\Big|_0^{2.0 \text{ s}} = (7.5 \text{ rad/s}^2)(2.5 \text{ N} \bullet \text{m})(2.0 \text{ s})^2 = \underline{75 \text{ J}}$$

13-19. The power is that required to increase the potential energy of the tractor per unit time. It is given by

$$P = \frac{\Delta U}{\Delta t} = mg\frac{\Delta h}{\Delta t} = mgv_y = mgv\sin\theta$$

$$= (4500 \text{ kg})(9.81 \text{ m/s}^2)(4.0 \text{ m/s})\sin(\tan^{-1}\tfrac{1}{3}) = \underline{5.6 \times 10^4 \text{ W}}$$

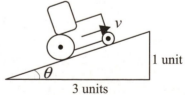

1 unit

3 units

The torque exerted by the engine on the rear wheel is

$$\tau = \frac{P}{\omega} = \frac{P}{v/r} = \frac{rP}{v} = \frac{(0.80 \text{ m})(5.6 \times 10^4 \text{ W})}{4.0 \text{ m/s}} = \underline{1.1 \times 10^4 \text{ N} \bullet \text{m}}$$

Note that the speed of the tractor up the incline is used rather than the speed in the vertical direction.

13-21. As the stick rotates about the pivot from the horizontal position to the vertical position, the center of mass of the meter stick is lowered by 0.1 m. Define the gravitational potential energy at this lowest position to be equal to zero. By applying the principle of the conservation of energy as the potential energy is converted into rotational energy, we find

$$mgh = \frac{1}{2}I\omega^2$$

The rotational inertia of the meter stick may be found using the parallel axis theorem.

$$I = I_m + md^2 = \frac{1}{12}ml^2 + m(0.10 \text{ m})^2 l^2 = 0.093ml^2$$

The angular speed may then be determined using the information given.

$$\omega^2 = \frac{2mgh}{I}$$

$$\omega = \sqrt{\frac{2mgh}{I}} = \sqrt{\frac{2mgh}{0.093ml^2}} = \sqrt{\frac{2(9.81 \text{ m/s}^2)(0.10 \text{ m})}{0.093(1.0 \text{ m})^2}} = \underline{4.6 \text{ rad/s}}$$

13-23. (a) The work done by the engine torque on the gear box as the engine makes 2.58 rev is

$$W = \tau\theta = (441 \text{ N}\cdot\text{m})(2.58 \text{ rev})\left(\frac{2\pi \text{ rad}}{1 \text{ rev}}\right) = 7150 \text{ J}$$

If all of this work is transferred to the wheels, then a torque τ' will be applied to the wheels and cause them to rotate one revolution (2π rad).

$$\tau' = \frac{W}{\theta'} = \frac{7150 \text{ J}}{2\pi \text{ rad}} = \underline{1140 \text{ N}\cdot\text{m}}$$

(b) The torque on the wheels is equal to the torque due to the force acting at the area of contact between the wheels and the ground. The ground force acts at a distance R, the radius of the wheels.

$$F = \frac{\tau'}{R} = \frac{1140 \text{ N}\cdot\text{m}}{0.30 \text{ m}} = \underline{3800 \text{ N}}$$

The acceleration that results from the ground forces on the automobile is

$$a = \frac{F}{M} = \frac{3800 \text{ N}}{1770 \text{ kg}} = \underline{2.1 \text{ m/s}^2}$$

13-25. The downward force **F** of the piston causes a torque on the crankshaft at a distance l from the pivot point on the crankshaft. The torque is $\tau = Fl$. Therefore, the force may be determined.

$$F = \frac{\tau}{l} = \frac{31 \text{ N}\cdot\text{m}}{0.038 \text{ m}} = \underline{820 \text{ N}}$$

13-27. The magnitude of angular acceleration is constant as the bridge opens to 45° and it decelerates at the same rate as it continues to open to 90°. Since the angular acceleration is constant as it opens to 45° in a 30 s interval, the angular displacement as a function of time is given by

$$\phi = \omega_0 t + \frac{1}{2}\alpha t^2$$

and the angular acceleration is

$$\phi = \frac{1}{2}\alpha t^2$$

$$\alpha = \frac{2\phi}{t^2} = \frac{2\left(\frac{\pi}{4}\text{rad}\right)}{(30 \text{ s})^2} = 1.7 \times 10^{-3} \text{ rad/s}^2$$

Since the bridge is treated as uniform rod with an axis through its center, $I = \frac{1}{12}Ml^2$. The torque is

$$\tau = I\alpha = \frac{1}{12}Ml^2\alpha = \frac{1}{12}(300\text{ t})\left(\frac{1000\text{ kg}}{1\text{ t}}\right)(25\text{ m})^2(1.7 \times 10^{-3}\text{ rad/s}^2) = \underline{2.7 \times 10^4\text{ N} \cdot \text{m}}$$

13-29.   Apply equation (13.24), to solve for g, the acceleration due to gravity.

$$a = \frac{m_2 - m_1}{m_1 + m_2 + (I/R^2)}g \qquad (13.24)$$

$$g = a\frac{m_1 + m_2 + (I/R^2)}{m_2 - m_1}$$

The acceleration is found by using the kinematic relation,

$$x = \frac{1}{2}at^2$$

$$a = \frac{2x}{t^2} = \frac{2(1.6\text{ m})}{(8.0\text{ s})^2} = 5.0 \times 10^{-2}\text{ m/s}^2$$

The rotational inertia of the disk is $I = \frac{1}{2}MR^2$, where M is the mass of the disk and R is its radius.

$$g = a\frac{m_1 + m_2 + \left(\frac{1}{2}M\right)}{m_2 - m_1} = 5.0 \times 10^{-2}\text{ m/s}^2\left(\frac{(0.4500\text{ kg}) + (0.4550\text{ kg}) + \frac{1}{2}(0.120\text{ kg})}{0.4550\text{ kg} - 0.4500\text{ kg}}\right) = \underline{9.7\text{ m/s}^2}$$

13-31.   The moment of inertia for a cylinder for an axis through its center is $I = \frac{1}{2}MR^2$. To find the moment for a rotation about a point on its rim, use the parallel axis theorem to find

$$I_0 = \frac{1}{2}MR^2 + MR^2 = \frac{3}{2}MR^2$$

The torque on the cylinder as it rolls down the incline is

$$\tau = I_0\alpha = \frac{3}{2}MR^2\alpha = MgR\sin\theta$$

$$a = R\alpha = R\frac{MgR\sin\theta}{\frac{3}{2}MR^2} = \underline{\frac{2}{3}g\sin\theta}$$

13-33.   The sphere is in contact with a plane that is inclined 20° as shown. The torque on the sphere due to the contact force **F** that is directed up the plane is

$$\tau = I_0\alpha = FR = MgR\sin\theta \qquad (i)$$

The moment of inertia for a sphere for an axis through its center is

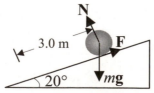

$I = \frac{2}{5}MR^2$. To find the moment for a rotation about a point on the sphere, use the parallel axis theorem to find

$$I_0 = \frac{2}{5}MR^2 + MR^2 = \frac{7}{5}MR^2$$

Then, using equation (i)

$$a = R\alpha = R\frac{MgR\sin\theta}{I_0} = \frac{MgR^2\sin\theta}{\frac{7}{5}MR^2} = \frac{5}{7}g\sin 20° = 0.244g \qquad (ii)$$

The problem also states that the sphere moves $s = 3.0$ m in $t = 1.6$ s. This information can be used to find the acceleration of the sphere down the plane from the kinematic relation

$$x = \frac{1}{2}at^2$$

$$a = \frac{2x}{t^2} \qquad (iii)$$

Subsituting equation (iii) into (ii) and solving for the acceleration due to gravity g gives

$$a = 0.244g = \frac{2x}{t^2}$$

$$g = \frac{2x}{0.244t^2} = \frac{2(3.0\text{ m})}{0.244(1.6\text{ s})^2} = \underline{9.6\text{ m/s}^2}$$

13-37.　The sphere is in contact with a plane that is inclined 15°. The torque on the ball due to the contact force F that is directed up the plane is

$$\tau = I_0\alpha = FR = MgR\sin\theta$$

The moment of inertia for a sphere for an axis through its center is $I = \frac{2}{5}MR^2$. To find the moment for a rotation about a point on the sphere, use the parallel axis theorem to find

$$I_0 = \frac{2}{3}MR^2 + MR^2 = \frac{5}{3}MR^2$$

Then, using the torque equation,

$$\alpha = \frac{MgR\sin\theta}{I_0} = \frac{MgR\sin\theta}{\frac{5}{3}MR^2} = \frac{3g}{5R}(\sin 15°) = 0.155\frac{g}{R}$$

Use the rotational kinematic relation to find the angle ϕ through which the ball rolls.

$$\phi = \omega_0 t + \frac{1}{2}\alpha t^2 = \frac{1}{2}\alpha t^2 = \frac{1}{2}\left(0.155\frac{g}{R}\right)t^2$$

$$= \frac{1}{2}\left(\frac{(0.155)(9.81\text{ m/s}^2)}{\frac{1}{2}(0.23\text{ m})}\right)(4.0\text{ s})^2 = 106\text{ rad}\left(\frac{1\text{ rev}}{2\pi\text{ rad}}\right) = \underline{17\text{ rev}}$$

13-39.　The turntable is rotationally accelerated from rest to an angular speed

$$\omega = 33.3\frac{\text{rev}}{\text{min}}\left(\frac{1\text{ min}}{60\text{ s}}\right)\left(\frac{2\pi\text{ rad}}{1\text{ rev}}\right) = 3.48\text{ rad/s}$$

The torque is applied at the outer rim of the disk of radius R and mass M. The torque required is

$$\tau = I\alpha = I\left(\frac{\omega}{t}\right) = \frac{1}{2}MR^2\left(\frac{\omega}{t}\right) = \frac{1}{2}(1.2\text{ kg})(0.15\text{ m})^2\frac{3.48\text{ rad/s}}{2.0\text{ s}} = \underline{0.024\text{ N}\cdot\text{m}}$$

The tangential force applied at the rim is

$$F = \frac{\tau}{R} = \frac{0.024\text{ N}\cdot\text{m}}{0.15\text{ m}} = \underline{0.16\text{ N}}$$

13-41.   The maximum magnitude of the static friction force at the point of contact of the hoop with the plane is

$$f = \mu_s N = \mu_s Mg \cos \theta \qquad \text{(i)}$$

The moment of inertia for a hoop for an axis through its center is $I = MR^2$. To find the moment for a rotation about a point on its rim, use the parallel axis theorem to find

$$I_0 = MR^2 + MR^2 = 2MR^2$$

There is also a torque on the hoop about the point of contact of the hoop due to the component of the weight parallel to the inclined plane.

$$\tau = I_0 \alpha = 2MR^2 \alpha = MgR \sin \theta$$

$$a = R\alpha = R\frac{MgR \sin \theta}{2MR^2} = \frac{1}{2} g \sin \theta$$

Combining this result with equation (i) gives

$$f = \mu_s Mg \cos \theta = Ma = \frac{1}{2} Mg \sin \theta$$

$$2\mu_s \cos \theta = \sin \theta$$

$$\underline{\tan \theta = 2\mu_s}$$

13-43.   The car described in problem 12-63 has a mass $M = 1360$ kg and each of the four wheels has a mass of $m = 27.2$ kg. The radius of each wheel is $R = 0.381$ m. Let f represent the friction force at the point of contact between the wheels and the road. As a result, the torque on each wheel is $\tau = Rf$. The torque results in an angular acceleration α of the wheel given by $\tau = I\alpha$. Therefore,

$$f = \frac{I\alpha}{R^2} \qquad \text{(i)}$$

There are five horizontal forces acting on the car, force **F** pulling the car forward and four friction forces acting at the wheels, in the opposite direction. The sum of these forces acting on the car, applying the result from equation (i) and using the relation $a = \alpha R$, is

$$F - 4f = F - 4\left(\frac{Ia}{R^2}\right) = Ma$$

The rotational inertia for each wheel is $I = \frac{1}{2}mR^2$. Therefore,

$$F - 4\left(\frac{\frac{1}{2}mR^2 a}{R^2}\right) = Ma$$

$$F - 2ma = Ma$$

$$a = \frac{F}{M + 2m} = \frac{4000 \text{ N}}{1360 \text{ kg} + 2(27.2 \text{ kg})} = \underline{2.83 \text{ m/s}^2}$$

If the rotational inertia of the wheels is neglected, the acceleration is

$$a = \frac{F}{M} = \frac{4000 \text{ N}}{1360 \text{ kg}} = \underline{2.94 \text{ m/s}^2}$$

The percent difference between these two values of the acceleration is

$$\left|\frac{2.94 - 2.83}{2.83}\right| \times 100 \% = \underline{3.89 \%}$$

13-47. The orbital angular momentum is given by $L = mvr$. The orbital radius and the orbital period are given for the moon. The average speed of the moon can be determined by dividing the circumference of the moon's orbit by the period.

$$L = mvr = m\left(\frac{2\pi r}{t}\right)r = \frac{2\pi mr^2}{t} = \frac{2\pi(7.35 \times 10^{22} \text{ kg})(3.8 \times 10^8 \text{ m})^2}{27.3 \text{ d}\left(\frac{24 \text{ h}}{1 \text{ d}}\right)\left(\frac{3600 \text{ s}}{1 \text{ h}}\right)} = \underline{2.8 \times 10^{34} \text{ kg} \cdot \text{m}^2/\text{s}}$$

13-51. The mass of the electron is $m = 9.11 \times 10^{-31}$ kg. The orbital angular momentum is $L = mvr$. For the first radius and velocity data given,

$$L_1 = mvr = (9.11 \times 10^{-31} \text{ kg})(2.18 \times 10^6 \text{ m/s})(0.529 \times 10^{-10} \text{ m}) = \underline{1.05 \times 10^{-34} \text{ kg} \cdot \text{m}^2/\text{s}}$$

For the other two "orbits,"

$$L_2 = mvr = (9.11 \times 10^{-31} \text{ kg})(1.09 \times 10^6 \text{ m/s})(2.12 \times 10^{-10} \text{ m}) = \underline{2.11 \times 10^{-34} \text{ kg} \cdot \text{m}^2/\text{s}}$$

$$L_3 = mvr = (9.11 \times 10^{-31} \text{ kg})(7.27 \times 10^6 \text{ m/s})(4.76 \times 10^{-10} \text{ m}) = \underline{3.15 \times 10^{-34} \text{ kg} \cdot \text{m}^2/\text{s}}$$

In comparing the three results, we note that $L_2 = 2L_1$ and $L_3 = 3L_1$. One might conclude that the trend would continue and that the orbital angular momentum is quantized, $L = nL_1$.

13-55. The angular momentum is $L = I\omega$, where the moment of inertia for a solid sphere is $I = \frac{2}{5}mR^2$ for rotations about an axis that passes through the center of the sphere. The spin angular speed is then

$$\omega = \frac{L}{I} = \frac{L}{\frac{2}{5}MR^2} = \frac{5.3 \times 10^{-35} \text{kg} \cdot \text{m}^2/\text{s}}{\frac{2}{5}(1.67 \times 10^{-27} \text{ kg})(1.0 \times 10^{-15} \text{ m})^2} = \underline{7.9 \times 10^{22} \text{ rad/s}}$$

From the angular speed, the rotational kinetic energy and the linear speed at a point on the equator of the sphere can be found.

$$v = R\omega = (1 \times 10^{-15} \text{ m})(7.9 \times 10^{22} \text{ rad/s}) = \underline{7.9 \times 10^7 \text{ m/s}}$$

$$K_{Rot} = \frac{1}{2}I\omega^2 = \frac{1}{2}\left(\frac{2}{5}MR^2\right)\omega^2$$

$$= \frac{1}{2}\left[\frac{2}{5}(1.67 \times 10^{-27} \text{ kg})(1.0 \times 10^{-15} \text{ m})^2(7.9 \times 10^{22} \text{ rad/s})^2\right] = \underline{2.1 \times 10^{-12} \text{ J}}$$

Finally, from Einstein's theory of relativity (see Chapter 8 or 36), the relationship between an object's mass and its energy is

$$E_0 = mc^2 = (1.67 \times 10^{-27} \text{ kg})(3.0 \times 10^8 \text{ m/s})^2 = \underline{1.5 \times 10^{-10} \text{ J}}$$

13-59. The spin angular momentum of the Sun is

$$L_S = I\omega = \frac{1}{5}MR^2\omega$$

$$= \frac{1}{5}(1.99 \times 10^{30} \text{ kg})(6.96 \times 10^8 \text{ m})^2\left(\frac{2\pi \text{ rad}}{25 \text{ d}}\right)\left(\frac{1 \text{ d}}{24 \text{ h}}\right)\left(\frac{1 \text{ h}}{3600 \text{ s}}\right) = \underline{5.6 \times 10^{41} \text{ kg} \cdot \text{m}^2/\text{s}}$$

The angular momentum for the planets orbiting the Sun is the sum of the individual angular momenta of the planets. The masses, mean distance from the Sun, and period of the planets is given in Table 9.1.

$$L = \sum_i MR_i^2 \omega_i = (3.30 \times 10^{23} \text{ kg})(57.9 \times 10^9 \text{ m})^2 \left(\frac{2\pi \text{ rad}}{0.241 \text{ y}}\right)\left(\frac{1 \text{ y}}{3.156 \times 10^7 \text{ s}}\right) +$$

$$(4.87 \times 10^{24} \text{ kg})(108 \times 10^9 \text{ m})^2 \left(\frac{2\pi \text{ rad}}{0.615 \text{ y}}\right)\left(\frac{1 \text{ y}}{3.156 \times 10^7 \text{ s}}\right) +$$

$$(5.98 \times 10^{24} \text{ kg})(150 \times 10^9 \text{ m})^2 \left(\frac{2\pi \text{ rad}}{1.000 \text{ y}}\right)\left(\frac{1 \text{ y}}{3.156 \times 10^7 \text{ s}}\right) +$$

$$(6.42 \times 10^{23} \text{ kg})(228 \times 10^9 \text{ m})^2 \left(\frac{2\pi \text{ rad}}{1.88 \text{ y}}\right)\left(\frac{1 \text{ y}}{3.156 \times 10^7 \text{ s}}\right) +$$

$$(1.90 \times 10^{27} \text{ kg})(778 \times 10^9 \text{ m})^2 \left(\frac{2\pi \text{ rad}}{11.9 \text{ y}}\right)\left(\frac{1 \text{ y}}{3.156 \times 10^7 \text{ s}}\right) +$$

$$(5.66 \times 10^{26} \text{ kg})(1430 \times 10^9 \text{ m})^2 \left(\frac{2\pi \text{ rad}}{29.5 \text{ y}}\right)\left(\frac{1 \text{ y}}{3.156 \times 10^7 \text{ s}}\right) +$$

$$(8.70 \times 10^{25} \text{ kg})(2870 \times 10^9 \text{ m})^2 \left(\frac{2\pi \text{ rad}}{84.0 \text{ y}}\right)\left(\frac{1 \text{ y}}{3.156 \times 10^7 \text{ s}}\right) +$$

$$(1.03 \times 10^{26} \text{ kg})(4500 \times 10^9 \text{ m})^2 \left(\frac{2\pi \text{ rad}}{165 \text{ y}}\right)\left(\frac{1 \text{ y}}{3.156 \times 10^7 \text{ s}}\right) +$$

$$(1.50 \times 10^{22} \text{ kg})(5890 \times 10^9 \text{ m})^2 \left(\frac{2\pi \text{ rad}}{248 \text{ y}}\right)\left(\frac{1 \text{ y}}{3.156 \times 10^7 \text{ s}}\right) = \underline{3.14 \times 10^{43} \text{ kg} \cdot \text{m}^2/\text{s}}$$

The percentage of the total angular mometum of the solar system (including only the Sun and planets) that is in the rotational motion of the Sun is

$$\text{Percentage} = 100 \times \frac{L_S}{L + L_S} = 100 \times \frac{5.61 \times 10^{41} \text{ kg} \cdot \text{m}^2/\text{s}}{3.13 \times 10^{43} \text{ kg} \cdot \text{m}^2/\text{s} + 5.61 \times 10^{41} \text{ kg} \cdot \text{m}^2/\text{s}} = \underline{1.8 \text{ \%}}$$

13-61.   The initial angular momentum is $L_1 = I_1 \omega_1 = \frac{1}{2} MR^2 \omega_1$

Because the child jumps onto the outer edge from along a radial line, she doesn't apply a torque that would change the angular velocity of the merry-go-round, but the addition of her mass to the merry-go-round does change the angular velocity because the angular momentum must be conserved.

$$L_1 = L_2 = I_2 \omega_2 = \left(\frac{1}{2} MR^2 + mR^2\right) \omega_2$$

$$\frac{1}{2} MR^2 \omega_1 = \left(\frac{1}{2} MR^2 + mR^2\right) \omega_2$$

$$\omega_2 = \frac{\frac{1}{2} MR^2}{\frac{1}{2} MR^2 + mR^2} \omega_1 = \frac{\frac{1}{2} M}{\frac{1}{2} M + m} \omega_1 = \frac{\frac{1}{2}(20 \text{ kg})}{\frac{1}{2}(20 \text{ kg}) + (25 \text{ kg})}(2.0 \text{ rad/s}) = \underline{0.57 \text{ rad/s}}$$

The child then applies a torque at the outer edge to increase the angular speed to 2.0 rad/s. In ther process, the angular momentum is increased. As the child walks from the outer edge toward the axis of rotation, the angular velocity must further increase to maintain a constant angular momentum.

$$L_3 = L_4$$

$$\left(\frac{1}{2}MR^2 + mR^2\right)\omega_1 = \left(\frac{1}{2}MR^2 + mr^2\right)\omega_3$$

$$\omega_3 = \frac{\frac{1}{2}MR^2 + mR^2}{\frac{1}{2}MR^2 + mr^2}\omega_1 = \left[\frac{\frac{1}{2}(20\text{ kg})(1.5\text{ m})^2 + (25\text{ kg})(1.5\text{ m})^2}{\frac{1}{2}(20\text{ kg})(1.5\text{ m})^2 + (25\text{ kg})(0.5\text{ m})^2}\right](2.0\text{ rad/s}) = \underline{5.5\text{ rad/s}}$$

13-63. The initial angular momentum is $L_1 = (I_0 + I)\omega_1$, where I_0 is the rotational inertia of the professor and stool and $I = 2mr^2$ is the rotational inertia of the two masses held at a distance r from the rotational axis. When the masses and arms are pulled in both rotational inertias decrease, so the angular speed must increase to maintain a constant angular momentum.

$$L_1 = L_2$$

$$(I_0 + I)\omega_1 = (I_0' + I')\omega_2$$

$$\omega_2 = \frac{I_0 + I}{I_0' + I'}\omega_1 = \frac{(6.0\text{kg}\cdot\text{m}^2) + 2(10\text{ kg})(1.0\text{ m})}{(4.0\text{kg}\cdot\text{m}^2) + 2(10\text{ kg})(0.10\text{ m})}\left(0.50\frac{\text{rev}}{\text{s}}\right) = \underline{2.2\text{ rev/s (or 14 rad/s)}}$$

13-67. The two satellites are in uniform circular motion as they orbit the Earth. They are held in orbit by the Earth's gravity, which provides the centripetal force.

$$G\frac{Mm}{R^2} = \frac{mv^2}{R}$$

Therefore, at a given radius, the speed will be

$$v^2 = G\frac{M_E}{R}$$

$$v = \sqrt{\frac{GM_E}{R}} \qquad \text{which gives} \qquad v_1 = \sqrt{\frac{GM_E}{r_1}} \quad \text{and} \quad v_2 = \sqrt{\frac{GM_E}{r_2}}$$

For the two satellites, since $r_1 < r_2$, $v_1 > v_2$. The satellite closer to the Earth has the greater speed.

The angular momentum of each is

$$L_1 = mv_1 r_1 = mr_1\sqrt{\frac{GM_E}{r_1}} = m\sqrt{r_1 GM_E}$$

$$L_2 = mv_2 r_2 = mr_2\sqrt{\frac{GM_E}{r_2}} = m\sqrt{r_2 GM_E}$$

Since $r_1 < r_2$, $L_2 > L_1$. The satellite further from Earth has the greater angular momentum.

13-71. The total angular momentum of the boat and woman before the woman turns is equal to zero, therefore as she turns, it must remain zero. Let the woman's angular momentum be $L_w = I_w \omega_w$ and the angular momentum of the boat be $L_b = I_b \omega_b$. Then,

$$0 = L_w + L_b = I_w\omega_w + I_b\omega_b$$

If the above equation is multiplied by Δt, the elapsed time for the woman's rotation, one obtains the angular displacement of the boat.

$$I_w \omega_w \Delta t + I_b \omega_b \Delta t = 0$$

$$I_w \phi_w + I_b \phi_b = 0$$

$$\phi_b = -\frac{I_w \phi_w}{I_b} = -\frac{(0.80 \text{ kg} \cdot \text{m}^2)(180°)}{20 \text{ kg} \cdot \text{m}^2} = \underline{-7°}$$

The rowboat rotates seven degrees in the direction opposite to that of the woman.

13-91. (a) From the information given, a linear equation for the torque due to the spring as a function of angle in radians can be written.

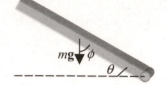

$$\tau_S = 2.00 \times 10^3 \text{ N} \cdot \text{m} - \frac{2.00 \times 10^3 \text{ N} \cdot \text{m} - 0.30 \times 10^3 \text{ N} \cdot \text{m}}{\frac{\pi}{2} \text{ rad}} \theta$$

$$= 2.00 \times 10^3 \text{ N} \cdot \text{m} - (1.08 \times 10^3 \text{ N} \cdot \text{m}/\text{rad})\theta$$

The angle at which this torque is equal to the torque due to the force of gravity τ_G on the hatch can be found. Consider the drawing, $\tau_G = \frac{1}{2} lmg \cos\theta$

Then,

$$\frac{1}{2} lm \, g\cos\theta = 2.00 \times 10^3 \text{ N} \cdot \text{m} - (1.08 \times 10^3 \text{ N} \cdot \text{m}/\text{rad})\theta$$

$$\frac{1}{2}(1.2 \text{ m})(400 \text{ kg})(9.81 \text{ m/s}^2) \cos\theta = 2.00 \times 10^3 \text{ N} \cdot \text{m} - (1.08 \times 10^3 \text{ N} \cdot \text{m}/\text{rad})\theta$$

$$(1.08 \times 10^3 \text{ N} \cdot \text{m}/\text{rad})\theta + (2.35 \times 10^3 \text{ N} \cdot \text{m})\cos\theta - 2.00 \times 10^3 \text{ N} \cdot \text{m} = 0$$

To determine the angle, one must use the graphical method or trial and error. The graph on the left below is a graph of the left hand side of the above equation versus the angle $0 \le \theta \le \pi/2$. From the graph on the right, the value of the function is zero at $\underline{\theta = 1.33 \text{ radians} = 76.2°}$.

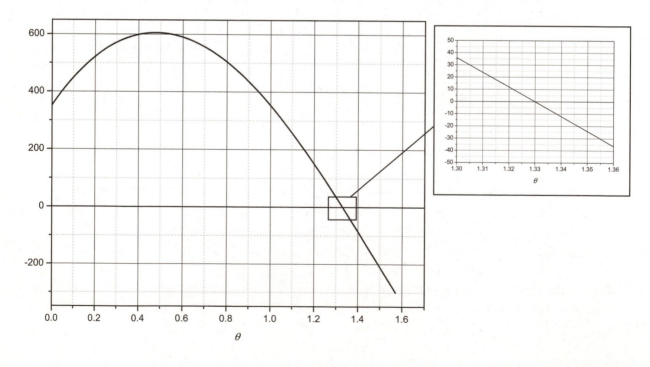

(b) In the closed position, $\theta = 0$, and the sum of the torques will be;

$$(1.08 \times 10^3 \text{ N} \cdot \text{m} / \text{rad})\theta + (2.35 \times 10^3 \text{ N} \cdot \text{m})\cos \theta - 2.00 \times 10^3 \text{ N} \cdot \text{m} - F(1.2 \text{ m}) = 0$$

$$F = \frac{(2.35 \times 10^3 \text{ N} \cdot \text{m})\cos 0 - 2.00 \times 10^3 \text{ N} \cdot \text{m}}{1.2 \text{ m}} = \underline{290 \text{ N}}$$

In the upright position, the force required is

$$(1.08 \times 10^3 \text{ N} \cdot \text{m} / \text{rad})\theta + (2.35 \times 10^3 \text{ N} \cdot \text{m})\cos \theta - 2.00 \times 10^3 \text{ N} \cdot \text{m} - F(1.2 \text{ m}) = 0$$

$$F = \frac{(1.08 \times 10^3 \text{ N} \cdot \text{m} / \text{rad})\frac{\pi}{2} + (2.35 \times 10^3 \text{ N} \cdot \text{m})\cos\left(\frac{\pi}{2}\right) - 2.00 \times 10^3 \text{ N} \cdot \text{m}}{1.2 \text{ m}} = \underline{-250 \text{ N}}$$

The minus sign indicates that the direction of the force, and the torque, has changed from the earlier case, as one would expect.

CHAPTER 14   STATICS AND ELASTICITY

14-1.   Three forces act in the rope-bucket system shown in the drawing. The forces are the tension **T** in the rope, the force **F** the man exerts on the bucket, and the weight **W** of the bucket. The angle θ is found from the length of the rope, 20 m, and the distance from the wall, 2.0 m:

$\theta = \sin^{-1}(2.0/20)$

The sum of the horizontal forces acting on the bucket is

$F - T \sin \theta = 0$

In the vertical direction, the sum is

$T \cos \theta - W = 0$ or $T = (W/\cos \theta)$

Combining these two equations, we can solve for the magnitude of F in terms of the weight of the bucket.

$$F - \left(\frac{W}{\cos\theta}\right)\sin\theta = 0$$

$$F = W\frac{\sin\theta}{\cos\theta} = W\tan\theta = mg\tan\theta$$

$$= (600 \text{ kg})(9.8 \text{ m/s}^2)\tan\left[\sin^{-1}\left(\frac{2.0}{20}\right)\right] = \underline{590 \text{ N}}$$

14-5.   Consider the drawing to the right. Initially the rope is taut between the car and the tree. The driver pushes on the rope with the force **F** shown at the mid-point and the tension in the rope is **T**. Because the angle at the top of the triangle is given as 170°, the two angles labeled θ are each 5° because the three angles of a triangle must add to 180°. Since the rope is in stable equilibrium, the sum of the three forces shown must be equal to zero newtons. The x components of the two tension forces are equal in magnitude, but oppositely directed, so

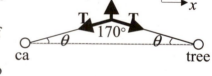

$T_x - T_x = (T \cos \theta) - (T \cos \theta) = 0$

In the y direction, the sum of the forces is

$F - T_y - T_y = F - (T \sin \theta) - (T \sin \theta) = F - 2T \sin \theta = 0$

Solving for T gives,

$$T = \frac{F}{2\sin\theta} = \frac{900 \text{ N}}{2 \sin 5°} = 5163 \text{ N} = \underline{5200 \text{ N}}$$

As long as the rope doesn't break under this tension, this is a highly effective way of pulling the car out of the mud.

14-9. Assuming the mass of the balance arm is neglected along with the masses of the pans, this problem may be solved by recognizing that the upward force **F** at the pivot does not contribute when considering the torques on the system. The moment arm for the weight of the small standard mass $m\mathbf{g}$ is 49 times longer than the moment arm of the weight of the bag of sugar, so we expect that the mass of the bag of sugar will be 49 times greater than the mass m. The sum of the torques is

$Mg(0.01\text{ m}) - mg(0.49\text{ m}) = 0$

$M(0.01\text{ m}) = m(0.49\text{ m})$

$M = 49m = 49(0.12\text{ kg}) = \underline{5.88\text{ kg}}$

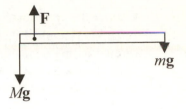

14-11. The forces acting on the mast are illustrated in the drawing. The weight of the mast is to be neglected. The mast is pivoted at the bottom. Consider the torques acting on the mast.

$(T_2 \sin 45°)(10\text{ m}) - (T_1 \sin 30°)(10\text{ m}) = 0$

$T_2 = T_1 \dfrac{\sin 30°}{\sin 45°} = (5.0 \times 10^3\text{ N})\dfrac{\sin 30°}{\sin 45°} = \underline{3500\text{ N}}$

The force that *the* mass exerts on the boat is equal in magnitude to the normal force the boat exerts on the mast. The normal force can be found by summing the forces in the y direction.

$N - T_1 \cos 30° - T_2 \cos 45° = 0$

$N = T_1 \cos 30° + T_2 \cos 45° = (5.0 \times 10^3\text{ N})\cos 30° + (3.5 \times 10^3\text{ N})\cos 45° = \underline{6800\text{ N}}$

14-13. In lifting the lid, you can apply the force **F** vertically as shown or you could imagine applying it at some other angle. To rotate the lid to an angle θ requires a torque to be applied with the lid held at one end at the hinges. The maximum torque occurs when the force **F** is applied in the direction perpendicular to the moment arm that is the lid. In this case, the sum of the torques acting on the lid is

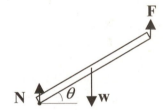

$FL - w\dfrac{L}{2}\cos\theta = 0$

$F = \dfrac{1}{2}w\cos\theta$

For $\theta = 30°$, $F = (0.5)(12\text{ kg})(9.81\text{ m/s}^2)\cos 30° = \underline{51\text{ N}}$
For $\theta = 60°$, $F = (0.5)(12\text{ kg})(9.81\text{ m/s}^2)\cos 60° = \underline{29\text{ N}}$

14-15. Choose the pivot location to be 1.0 m from the bolted end of the board where the upward support force $\mathbf{F}_{up}$ is applied. Let $\mathbf{F}_B$ represent the downward force at the bolted end. The sum of the torques acting on the board is

$$F_B(1.0 \text{ m}) - w_{board}(0.5 \text{ m}) - w_{diver}(2.0 \text{ m}) = 0$$

This equation may be then solved for F_B.

$$F_B = \frac{w_{board}(0.5 \text{ m}) + w_{diver}(2.0 \text{ m})}{1.0 \text{ m}}$$

$$= \frac{(50 \text{ kg})(9.81 \text{ m/s}^2)(0.5 \text{ m}) + (60 \text{ kg})(9.81 \text{ m/s}^2)(2.0 \text{ m})}{1.0 \text{ m}} = \underline{1420 \text{ N}}$$

The sum of the forces in the vertical direction is

$$F_{up} - F_B - w_{board} - w_{diver} = 0$$

and

$$F_{up} = F_B + w_{board} + w_{diver}$$
$$= 1420 \text{ N} + (50 \text{ kg})(9.81 \text{ m/s}^2) + (60 \text{ kg})(9.81 \text{ m/s}^2) = \underline{2500 \text{ N}}$$

14-19. The weight $\mathbf{w}$ of the locomotive is equal to the sum of all the downward forces of the wheels on the track.

$$\mathbf{w} = 2(74 \text{ kN}) + 4(117 \text{ kN}) + 109 \text{ kN} + 160 \text{ kN} + 153 \text{ kN} = 1.038 \times 10^6 \text{ N}$$

The downward forces of locomotive at the wheels on the track are equal in magnitude, but oppositely directed to the normal forces of the track on the locomotive. The sum of the torques acting on the train must equal zero N • m, if the train is in rotational equilibrium. Let x be the distance of the center of mass from the front wheel of the locomotive. The sum of the torques with the front wheel as the pivot point is equal to the weight of the locomotive times the moment arm x.

$$wx = (74 \text{ kN})(17.1 \text{ m}) + (74 \text{ kN})(15.4 \text{ m}) + (109 \text{ kN})(13.0 \text{ m}) + (160 \text{ kN})(10.7 \text{ m}) +$$
$$(153 \text{ kN})(8.4 \text{ m}) + (117 \text{ kN})(4.8 \text{ m}) + (117 \text{ kN})(3.1 \text{ m}) + (117 \text{ kN})(1.7 \text{ m})$$
$$= 7942 \text{ kN} \cdot \text{m}$$

$$x = (7.942 \times 10^6 \text{ N} \cdot \text{m})/(1.038 \times 10^6 \text{ N}) = \underline{7.65 \text{ m}}$$

$$F_A = 98 \text{ N at } 63° \text{ relative to the } +x \text{ axis and } F_B = 98 \text{ N at } 117° \text{ relative to the } +x \text{ axis}$$

14-21. The forces your fingers apply to the box are the friction force $\mathbf{f}_1$ and a normal force $\mathbf{N}_1$; and your thumb exerts similar forces as shown. The weight of the box is $\mathbf{w}$. If the box is being held in equilibrium, then the sum of the forces in the vertical direction is $f_2 - N_1 - w = 0$ which gives $f_2 = N_1 + w$ (i)

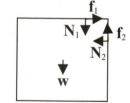

In the horizontal direction,

$f_1 - N_2 = 0$ which gives $f_1 = N_2$ (ii)

From the relationship between the friction force and the normal force, we have

$$f_1 = \mu N_1 = N_2 \qquad \text{(iii)}$$

Substituting the result from (iii) into (i) gives

$N_1 + w = f_2 = \mu N_2 = \mu^2 N_1$ from which μ can be determined.

$$\mu = \sqrt{\frac{N_1 + w}{N_1}}$$

Since the weight is always greater than zero N, μ will always be greater than 1. In the limit of the weight going to zero N, μ goes to 1.

14-23. Consider the drawing of the situation. The thin line drawn from the center of the wheel to the pivot point P represents the moment arm, which has a length equal to the radius of the wheel, R. The horizontal and vertical components of the normal force at the pivot point are not shown to avoid a cluttered drawing. Also, the wheel as it begins to rotate about the pivot point will lose contact with the ground; and it isn't necessary to consider the contact force at the point of contact with the ground.

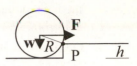

For the wheel to begin to roll over the step, the clockwise torque due to $\mathbf{F}$ must, however, just exceed the counterclockwise torque due to the weight. Let θ represent the angle between the horizontally applied force $\mathbf{F}$ and the moment arm. The critical value of this force may be found by summing these torques about P.

$wR \cos \theta - FR \sin \theta = 0$

From the drawing, the $\sin \theta$ and $\cos \theta$ terms can be written in terms of the radius of the wheel and the height of the step, h. Therefore, we can solve for F.

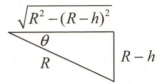

$$F = \frac{w}{\tan \theta} = \frac{Mg}{\tan \theta} = Mg\frac{\sqrt{R^2 - (R-h)^2}}{(R-h)}$$

The smallest force that one could apply would be the one applied perpendicular to the moment arm. In that case, the sum of torques is

$wR \cos \theta - FR = 0$

and

$$F = \frac{w}{\cos\theta} = \frac{Mg}{\cos\theta} = Mg\frac{R}{\sqrt{R^2 - (R-h)^2}}$$

14-25. (a) The base of the mast is hinged, so that will be the pivot point. The ropes make an angle θ with respect to the mast. We can find θ from the triangle formed by the rope, the mast, and the base.

$\theta = \tan^{-1}\left(\frac{1.35 \text{ m}}{10 \text{ m}}\right) = 7.7°$

The sum of the torques on the mast is

$F(5.0 \text{ m}) + T_2(\sin \theta)(10 \text{ m}) - T_1(\sin \theta)(10 \text{ m}) = 0$

The excess tension is $T_1 - T_2 = \dfrac{F(5.0 \text{ m})}{(10 \text{ m})\sin \theta} = \dfrac{(2400 \text{ N})(5.0 \text{ m})}{(10 \text{ m})\sin 7.7°} = \underline{9.0 \times 10^3 \text{ N}}$

(b) When spreaders are used, the ropes make a new angle with the mast:

$\theta = \tan^{-1}\left(\frac{1.35 \text{ m}}{2.5 \text{ m}}\right) = 28°$

and the sum of the torques is now,

$F(5.0 \text{ m}) + T_2(\sin \theta)(10 \text{ m}) - T_1(\sin \theta)(10 \text{ m}) = 0$

The excess tension in this case is

$T_1 - T_2 = \dfrac{F(5.0 \text{ m})}{(10 \text{ m})\sin \theta} = \dfrac{(2400 \text{ N})(5.0 \text{ m})}{(10 \text{ m})\sin 28°} = \underline{2.6 \times 10^3 \text{ N}}$

Since the excess tension is reduced by using the spreader, this is the preferred arrangement.

14-27. The legs form an equilateral triangle at the ground. Each side of the triangle has length L, which is also the length of the legs. To determine the tension in each leg, the angle θ between line AB and line AC must be determined. For an equilateral triangle, the distance AB from a vertex to the center of the triangle is given by

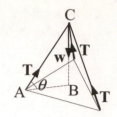

$$AB = \frac{2}{3}L\cos 30°$$

So, the angle θ may be found using

$$\cos\theta = \frac{AB}{AC} = \frac{\frac{2}{3}L\cos 30°}{L} = \frac{2}{3}\cos 30°$$

$$\theta = \cos^{-1}\left(\frac{2}{3}\cos 30°\right) = 54.7°$$

For static equilibrium, the sum of the forces in the vertical direction must be equal to zero newtons.

$$3T(\sin 54.7°) - Mg = 0$$

Therefore, the tension in each leg is

$$T = \frac{Mg}{3\sin 54.7°} = \underline{0.408Mg}$$

14-29. The sum of the forces acting on the disk in the vertical direction is

$$T\cos\theta - Mg = 0 \qquad \text{(i)}$$

and in the horizontal direction, it is $\qquad N - T\sin\theta = 0$ (ii)

To find the angle θ, the vertical distance h from the point of contact of the disk and the wall to the point where the string is attached to the wall is found using the Pythagorean theorem.

$$h^2 + R^2 = (L+R)^2$$

$$h = \sqrt{(L+R)^2 - R^2}$$

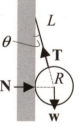

Then, $\cos\theta = \dfrac{\sqrt{(L+R)^2 - R^2}}{L+R}$ and $\sin\theta = \dfrac{R}{L+R}$

From equation (i), the tension is found.

$$T\frac{\sqrt{(L+R)^2 - R^2}}{L+R} - Mg = 0$$

$$T = Mg\left(\frac{L+R}{\sqrt{(L+R)^2 - R^2}}\right)$$

From equation (ii), the normal force may then be found.

$$N - Mg\left(\frac{L+R}{\sqrt{(L+R)^2 - R^2}}\right)\frac{R}{L+R} = 0$$

$$N = \frac{MgR}{\sqrt{(L+R)^2 - R^2}}$$

14-33. (a) The box tilted upward has a pivot axis that is the remaining bottom edge in contact with the floor. The center of mass moves along the arc of a circle of radius

$$R = \sqrt{(0.30 \text{ m})^2 + (0.60 \text{ m})^2} = 0.67 \text{ m}$$

The angle between the bottom of the box and the radial line between a pivot point and the center of mass is θ_0 as the box is tilted at an angle θ with respect to the floor.

$$\theta_0 = \tan^{-1}\left(\frac{0.60 \text{ m}}{0.30 \text{ m}}\right) = 63.4°$$

The height of the center of mass of the box is
$$h = R\left[\sin(\theta_0 + \theta) - \sin\theta_0\right]$$
and the potential energy is
$$\begin{aligned}
U = mgh &= mgR\left[\sin(\theta_0 + \theta) - \sin\theta_0\right] \\
&= (80 \text{ kg})(9.81 \text{ m/s}^2)(0.67 \text{ m})[\sin(63.4° + \theta) - \sin(63.4°)] \\
&= 526 \text{ J }[\sin(63.4° + \theta) - \sin(63.4°)]
\end{aligned}$$
The following figure is a graph of the potential energy as a function of θ.

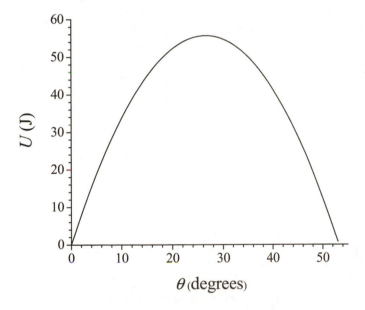

(b) The critical angle corresponds to the maximum of U (zero torque). This maximum is determined by $\sin(\theta + 63.4°) = 1$ or $\theta = \underline{26.6°}$

(c) The amount of work done on the box is equal to the potential energy at the maximum, assuming the potential energy was zero J when the box was at rest with the bottom completely in contact with the floor.

$$U = 526 \text{ J }[\sin(90.0°) - \sin(63.4°)] = \underline{56 \text{ J}}$$

14-37.   When the car is just about to topple, its normal force is acting at the edge of the outer wheel. The frictional force on the car is the centripetal force $F = mv^2/R$. For the car to balance, the torque about the *center of mass* must equal zero, that is,

$$Fh = \frac{1}{2}Nl$$

and to balance the vertical forces on the car, $N = mg$.

$$Fh = \frac{mv^2}{R}h = \frac{1}{2}Nl = \frac{1}{2}mgl$$

which gives $v = \sqrt{\frac{Rgl}{2h}} = \sqrt{\frac{(25 \text{ m})(9.81 \text{ m/s}^2)(1.5 \text{ m})}{2(0.60 \text{ m})}} = \underline{17.5 \text{ m/s}}$

14-39.   The bicycle is in stable equilibrium as long as the sum of the forces and torques acting on it are zero. However, it is ready to tip when the normal force on the back wheel N_{back} goes to zero newtons. The friction force is applied tangentially to the rim of the wheel at the radius R of the wheel. This friction force is equal in magnitude to the friction force of the road on the bicycle, which is

$f = \mu N_{front} = ma$

If the sum of the torques about the center of mass on the bicycle is zero, then

$N_{front}(d \sin\theta) = f(d \cos\theta)$

where d is the distance from the contact point to the center of mass. So, the maximum braking force that can be applied is

$f = N_{front} \tan\theta = mg \tan\theta = ma$

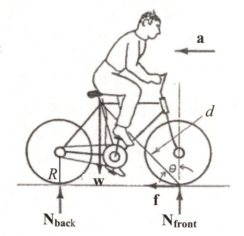

Therefore, $a = g\tan\theta = 9.81 \text{ m/s}^2 \left(\dfrac{0.70 \text{ m}}{0.95 \text{ m}}\right) = \underline{7.2 \text{ m/s}^2}$

since the center of mass is located 0.95 m above the road and 0.70 m behind the point of contact of the front wheel,

14-41.   (a) Let N_1 and N_2 represent the normal forces at the front and rear wheels, respectively. The weight is assumed to be distributed at the four wheels, so

$N_1 + N_2 = mg$ (i)

The friction forces are acting at both the front and rear wheels to accelerate the car. Applying Newton's second law of motion gives,

$f = ma = \mu N_1 + \mu N_2 = \mu (N_1 + N_2) = \mu mg$ (ii)

$a = \mu g = (0.90)(9.81 \text{ m/s}^2) = \underline{8.8 \text{ m/s}^2}$

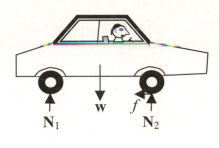

(b) Consider the sum of torques acting on the car using the center of mass as the pivot point.

$(1.5 \text{ m})N_1 + (0.6 \text{ m})(0.9)N_2 - (1.5 \text{ m})N_2 = 0$

This gives a relationship between N_1 and N_2.

$N_2 = 1.56N_1 = 1.56(mg - N_2)$

where the relationship from equation (i) has been used to eliminate N_1 to find N_2.

$N_2 = 0.61mg$

Apply Newton's second law to find the acceleration in this case.

$f = ma = \mu N_2 = \mu(0.61mg) = (0.90)(0.61)mg = 0.55mg$

$a = 0.55g = (0.55)(9.81 \text{ m/s}^2) = \underline{5.4 \text{ m/s}^2}$

(c) When the braking force is only on the rear wheels, the situation is similar to that in part (b). The sum of torques about the center of mass is

$(0.6 \text{ m})(0.9)N_1 + (1.5 \text{ m})N_1 - (1.5 \text{ m})N_2 = 0$

$N_1 = 0.75N_2 = 0.75(mg - N_1) = 0.43mg$

Apply Newton's second law to find the acceleration in this case.

$f = ma = \mu N_1 = \mu(0.43mg) = (0.90)(0.43)mg = 0.0.39mg$

$a = 0.39g = (0.39)(9.81 \text{ m/s}^2) = \underline{3.8 \text{ m/s}^2}$

14-43.   The bottom ball has two normal forces acting on it; and the top ball also has two normal forces acting on it as shown in the drawing. The sum of the horizontal forces is

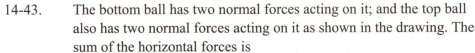

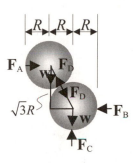

$F_A - F_B = 0$ which gives $F_A = F_B$

For the vertical forces,

$F_C - 2w = F_C - 2mg = 0$ which gives $F_C = \underline{2mg}$

Using the center of mass of the lower ball as a pivot point, the sum of the torques is

$mgR - \sqrt{3}RF_A = 0$

$F_A = F_B = \dfrac{mg}{\sqrt{3}}$

14-45.   The forces acting on the cards are shown in the free-body diagram. The frictional force at the top where the cards are in contact is negligible. Let θ represent the angle that the cards, of length L, make with respect to the vertical direction. The sum of the horizontal and vertical forces acting on one card is:

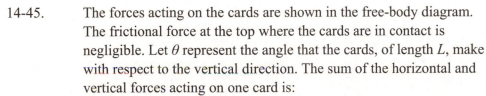

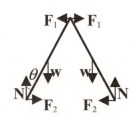

$F_2 - F_1 = 0$ or $F_1 = F_2$

and

$N - mg = 0$ or $N = mg$

For the maximum static friction force, $F_2 = \mu_s N = \mu_s mg = F_1$, at the maximum angle. The sum or torques about the edge of the card in contact with the floor is

$$\frac{1}{2}mgL\sin\theta_{max} - F_1 L\cos\theta_{max} = 0$$

$$\frac{1}{2}mgL\sin\theta_{max} - \mu_s mgL\cos\theta_{max} = 0$$

$$\frac{1}{2}\sin\theta_{max} - \mu_s\cos\theta_{max} = 0$$

$$\tan\theta_{max} = 2\mu_s \quad \text{and} \quad \theta_{max} = \underline{\tan^{-1}(2\mu_s)}$$

14-47.   The tension around each semi-circular portion of each flywheel is determined by the total friction force of the belt in contact with the flywheel. The friction force is at its maximum static value, $f_s = \mu_s N$. Consider the infinitesimal section of the belt shown in the force diagram.

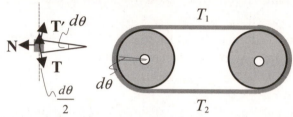

The magnitude of the horizontal component of each of the two tension forces $\mathbf{T}$ and $\mathbf{T'}$ is equal to

$$T\sin\frac{d\theta}{2} = T\frac{d\theta}{2} \quad \text{(small angle approximation)}$$

The equilibrium of the radial forces requires that

$$N - 2T\frac{d\theta}{2} = 0 \quad \text{or} \quad N = Td\theta$$

The friction force on the section is then given by $f_s = \mu_s N = \mu_s Td\theta$
The sum of the tangential forces on the section gives $dT = \mu_s Td\theta$ or

$$\frac{dT}{T} = \mu_s d\theta$$

Then, the total tension along the portion of belt in contact with the flywheel is found through integration.

$$\int_{T_1}^{T_2}\frac{dT}{T} = \mu_s\int_{\theta_1}^{\theta_2}d\theta$$

$$\ln T_2 - \ln T_1 = \mu_s(\theta_2 - \theta_1)$$

$$\text{or} \quad \theta_2 - \theta_1 = \frac{1}{\mu_s}\ln\frac{T_2}{T_1}$$

If we take the exponential of both sides of the equation above and rearrange, we get
$$T_2 = T_1 e^{\mu_s(\theta_2 - \theta_1)}$$

$\theta_2 - \theta_1 = \pi$, so $T_2 = T_1 e^{\mu_s\pi}$. The torque on the flywheel will be $\tau = R(T_2 - T_1)$. Therefore, the tensions may be written in terms of the torque.

$$\underline{T_1 = \frac{\tau}{R}\left(\frac{1}{e^{\mu_s\pi} - 1}\right)} \quad \text{and} \quad \underline{T_2 = \frac{\tau}{R}\left(\frac{1}{1 - e^{-\mu_s\pi}}\right)}$$

14-49. The arm is being held in static equilibrium, so the sum of the torques about the elbow (P) is equal to zero.

$(0.055 \text{ m})F_2 - (0.355 \text{ m})mg = 0$

Solving for the force F_2,

$$F_2 = \frac{(0.355 \text{ m})(25 \text{ kg})(9.81 \text{ m/s}^2)}{0.055 \text{ m}} = 1583 \text{ N} = \underline{1580 \text{ N}}$$

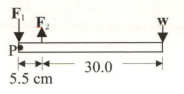

The sum of the vertical forces gives

$F_2 = F_1 + w$

So, $F_1 = F_2 - w = 1583 \text{ N} - (25 \text{ kg})(9.81 \text{ m/s}^2) = 1338 \text{ N} = \underline{1340 \text{ N}}$

14-51. The sum of the torques about the center of the cylinder gives

$F(0.25\text{m}) - T(0.040 \text{ m}) = 0$

$$F = \frac{(0.040 \text{ m})(2500 \text{ N})}{(0.25 \text{ m})} = \underline{400 \text{ N}}$$

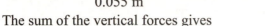

14-59. (a) Consider the windlass shown in Figure 14-30. Suppose we turn the handle 1 revolution, as indicated, by applying a torque τ. Then the right rope unwinds a distance $2\pi R_1$, and the left rope winds up a distance $2\pi R_2$. If the original cable length is $2H$, then the new cable length will be $2H + 2\pi(R_1 - R_2)$ and $\Delta L = L_2 - L_1 = 2\pi(R_1 - R_2)$. The load is moved upward a distance $h = \pi(R_1 - R_2)$, so the change in the potential energy of the load is $mg\pi(R_1 - R_2)$, which is equal to the rotational work done. Therefore, since $\tau(\Delta\phi) = 2\pi\tau$,

$$\tau = \frac{1}{2}mg(R_1 - R_2)$$

(b) If the length of the handle is l, then a sum of the torques about the rotational axis gives

$Fl - \tau = 0$

So, the tangential force F that must be applied to the handle is

$$F = \frac{1}{2}mg\frac{(R_1 - R_2)}{l}$$

(c) The mechanical advantage is

$$\frac{F'}{F} = \frac{2l}{R_1 - R_2}$$

14-61. The total mechanical advantage is a product of the mechanical advantage of each part of the system. The mechanical advantage is

$$MA = \frac{F'}{F} = \frac{x'}{x} = \left(\frac{16.0 \text{ cm}}{3.0 \text{ cm}}\right)\left(\frac{5.0 \text{ cm}}{3.0 \text{ cm}}\right) = \underline{8.9}$$

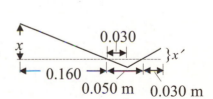

14-65. As seen in problem 57, the block and tackle has a mechanical advantage of 4. The total mechanical advantage of the system though is higher than this. As the horizontal bar is pulled down by an amount x, the top of the strings will move down by an amount x' to give a mechanical advantage of x/x'. Suppose we displace the horizontal bar downward by an amount x while constraining the string joint to remain stationary at the top. This will increase the string length by

$L' - L = x \sin 72.5° - x \cos 32.5° = 0.110x$

However, since the string length remains constant, the point will move down by this amount. Therefore, the mechanical advantage is $x/(0.11x)$. Thus, the total mechanical advantage is this system is

$$\text{TMA} = 4\left(\frac{x}{0.110x}\right) = \underline{36}$$

14-69. Ideal springs behave according to Hooke's law, $F = k\Delta x$, where a force F applied to one end of a spring with the other end fixed will proportionally stretch the spring by Δx. The constant of proportionality is the spring constant k. Similarly, in equation 14.18 that

$$\frac{\Delta L}{L} = \frac{1}{Y}\frac{F}{A}$$

$$\frac{F}{\Delta L} = \frac{AY}{L} = \frac{\pi(3.0 \times 10^{-4}\ \text{m})^2(22 \times 10^{10}\ \text{N/m}^2)}{1.8\ \text{m}} = \underline{3.5 \times 10^4\ \text{N/m}}$$

This is the same result that is found by substituting $F = 70$ N and $\Delta L = 2.0 \times 10^{-3}$ m (from problem 14-67.

14-73. Assuming the weight of the elevator is equally supported by the upward tension forces in the cables, the tension in each cable is

$$T = \frac{1}{3}Mg = \frac{1}{3}(1000\ \text{kg})(9.81\ \text{m/s}^2) = 3270\ \text{N}$$

The tension will stretch each steel cable by a distance

$$\Delta L = \frac{L}{Y}\frac{F}{A} = \left(\frac{300\ \text{m}}{22 \times 10^{10}\ \text{N/m}^2}\right)\frac{3270\ \text{N}}{\pi(5.0 \times 10^{-3}\ \text{m})^2} = \underline{0.057\ \text{m}}$$

14-75. The volume of the sphere is $V = \frac{4}{3}\pi R^3$. Take the derivative of the volume with respect to R.

$$\frac{dV}{dR} = \frac{d\left(\frac{4}{3}\pi R^3\right)}{dR} = 4\pi R^2$$

$$dV = 4\pi R^2\,dR$$

Then, the percent change in the volume is

$$\frac{dV}{V} = \frac{4\pi R^2\,dR}{\frac{4}{3}\pi R^3} = 3\frac{dR}{R}$$

This shows that the percent change in volume is equal to three times the percent change in the radius, for small changes in the radius. So, if the volume decreases by 0.10%, the radius changes by <u>0.033%</u>.

14-77. The slab is held by two iron bolts with a cross-sectional radius of 0.075 m. A downward shear force on each bolt due to one-half the weight of the slab deflects slightly downward by Δx. The shear force acts on each bolt at a distance, $h = 0.010$ m. The shear modulus of wrought iron is 7.7×10^{10} N/m^2.

$$\frac{\Delta x}{h} = \frac{1}{S}\frac{F}{A}$$

$$\Delta x = \frac{h}{S}\frac{F}{A} = \frac{(0.010\ \text{m})\frac{1}{2}(1200\ \text{kg})(9.81\ \text{m/s}^2)}{(7.7 \times 10^{10}\ \text{N/m}^2)\pi(0.0075\ \text{m})^2} = \underline{4.3 \times 10^{-6}\ \text{m}}$$

14-79. The breaking strength is equal to the product of the ultimate tensile strength and the cross-sectional area of the rope. Thus, if the breaking strength of the two spliced sections is equal, the diameter of the steel rope can be determined from

$$T_{U,\,\text{Steel}}\frac{\pi D_{\text{Steel}}^2}{4} = T_{U,\,\text{Nylon}}\frac{\pi D_{\text{Nylon}}^2}{4}$$

$$D_{\text{Steel}} = D_{\text{Nylon}}\sqrt{\frac{T_{U,\,\text{Nylon}}}{T_{U,\,\text{Steel}}}} = (1.3\ \text{cm})\sqrt{\frac{3.2 \times 10^8\ \text{N/m}^2}{2.0 \times 10^9\ \text{N/m}^2}} = \underline{0.52\ \text{cm}}$$

14-81. Consider the diagram shown. The sum of the vertical forces acting on the beam is $2T\sin\theta - w = 0$, from which

$$T = \frac{w}{2\sin\theta}$$

$$= \frac{mg}{2\sin\left[\cos^{-1}\left(\frac{1.0\ \text{m}}{3.0\ \text{m}}\right)\right]} = \frac{(8000\ \text{kg})(9.81\ \text{m/s}^2)}{2\sin 70.5°} = 4.2 \times 10^4\ \text{N}$$

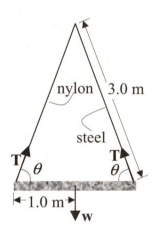

nylon   3.0 m

steel

T θ T θ

1.0 m

w

As a result of the tension, the two ropes stretch. The difference in extension of the two ropes is

$$\delta L = \Delta L_{\text{N}} - \Delta L_{\text{S}} = LT\left(\frac{1}{Y_{\text{N}}A_{\text{N}}} - \frac{1}{Y_{\text{S}}A_{\text{S}}}\right)$$

$$= (3.0\ \text{m})(4.2 \times 10^4\ \text{N})\left(\frac{1}{(0.36 \times 10^{10}\ \text{N/m}^2)\pi(0.0125\ \text{m})^2} - \frac{1}{(22 \times 10^{10}\ \text{N/m}^2)\pi(0.0032\ \text{m})^2}\right)$$

$$= 5.3 \times 10^{-2}\ \text{m} = 5.3\ \text{cm}$$

The angle the beam makes with the horizontal direction is

$$\theta = \tan^{-1}\frac{\delta L}{2.0\ \text{m}} = \tan^{-1}\left(\frac{0.053\ \text{m}}{2.0\ \text{m}}\right) = \underline{1.5°}$$

14-85.   Consider the drawing. An element dx is located a distance x from the center of mass through which the rotation axis passes. The centripetal force required to accelerate element dx is $dF = dmx\omega^2$. If A is the cross-sectional area of the meter stick, then $dm = \rho A\, dx$. Therefore, the tension on the stick will be the integral of the centripetal force for each element dx.

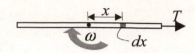

$$T = \int dF = \omega^2 \int x\,dm = \rho A\omega^2 \int_0^{l/2} x\,dx = \frac{1}{8}\rho A\omega^2$$

The maximum angular velocity is then

$$\omega_{max} = \sqrt{\frac{8}{\rho}\left(\frac{T}{A}\right)_{max}} = \sqrt{\frac{8}{7800 \text{ kg/m}^3}\left(3.8 \times 10^8 \text{ N/m}^2\right)} = \underline{624 \text{ rad/s}}$$

14-103.   The mass m of the rod is related to its volume V by $m = \rho V = \rho A L$, where ρ is the density, A is the cross-sectional area of the rod, and L is its length. The tensile stress is

$$\frac{F}{A} = \frac{w}{A} = \frac{mg}{A} = \frac{\rho A L g}{A} = \rho L g$$

The length at which the maximum tensile stress will be reached is

$$L = \frac{(F/A)_{max}}{\rho g} = \frac{3.8 \times 10^8 \text{ N/m}^2}{(7800 \text{ kg/m}^3)(9.81 \text{ m/s}^2)} = \underline{5.0 \times 10^3 \text{ m}}$$

CHAPTER 15   OSCILLATIONS

15-1.   (a) $A = \underline{3.0 \text{ m.}}$

angular frequency $\omega = 2.0$ rad/sec,

$\Rightarrow$ frequency $f = \dfrac{\omega}{2\pi} = \dfrac{2}{2\pi} = \dfrac{1}{\pi}$ Hz $= \underline{0.318 \text{ Hz}}$

period $T = 1/f = \pi = \underline{3.14 \text{ s.}}$

(b) $x = 0$ when $2.0t = \pi/2, 3\pi/2, 5\pi/2, \ldots$ The first time this occurs is $t = \dfrac{\pi}{4}$ s $= \underline{0.785 \text{ s.}}$

The turning points occur when $x = \pm A$. This happens when $2.0t = 0, \pi, 2\pi, 3\pi, \ldots$ The first time this occurs after $t = 0$ is $t = \dfrac{\pi}{2}$ s $= \underline{1.57 \text{ s.}}$

15-3.   (a) $f = \dfrac{1}{T} = \dfrac{1}{1.2}$ Hz $= \underline{0.83 \text{ Hz}}$

angular frequency $\omega = 2\pi f = \dfrac{\pi}{0.6}$ rad/s $= \underline{5.2 \text{ rad/s}}$

(b) $A = \underline{0.20 \text{ m.}}$

(c) The equation of motion is $x = A\cos\left(\dfrac{2\pi}{T}t\right) = 0.20\cos\left(\dfrac{\pi}{0.6}t\right)$. If $x = 0$, $\cos(\dfrac{\pi}{0.6}t) = 0$

$\Rightarrow \dfrac{\pi}{0.6}t = \dfrac{\pi}{2} \Rightarrow t = \underline{0.30 \text{ s.}}$ Because the period is 1.2 s, the particle will also pass through $x = 0$ at $t = 0.90$ s, 1.50 s, 2.10 s, etc. (In SHM the particle passes through the equilibrium position *twice* during each cycle.)

For $x = -0.10$ m, $-0.10 = 0.2\ \cos(\dfrac{\pi}{0.6}t)$

$\Rightarrow \quad \cos\dfrac{\pi}{0.6}t = -0.5 \quad \Rightarrow \quad \dfrac{\pi}{0.6}t = \dfrac{2\pi}{3}, \dfrac{4\pi}{3}, \dfrac{8\pi}{3}, \dfrac{10\pi}{3}, \cdots$

$\Rightarrow t = \underline{0.40 \text{ s, 0.80 s, 1.60 s, 2.00 s, } \ldots}$

(d) The speed of the motion is: $|v| = \left|\dfrac{dx}{dt}\right| = \dfrac{\pi}{0.6} \times 0.2 \times \left|\sin\dfrac{\pi t}{0.6}\right| = \dfrac{\pi}{3}\left|\sin\dfrac{\pi t}{0.6}\right|$

From (c) we know that the first time $x = 0$ corresponds to $t = 0.30$ s. At this time,

$|v| = \dfrac{\pi}{3}\left|\sin\dfrac{\pi(0.3)}{0.6}\right| = \dfrac{\pi}{3}$ m/s $= \underline{1.05 \text{ m/s.}}$

The first time $x = -0.1$ m corresponds to $t = 0.40$ s. At this time, $|v| = \dfrac{\pi}{3}\left|\sin\dfrac{\pi(0.4)}{0.6}\right| = \underline{0.91 \text{ m/s.}}$

15-5.   The equation of motion of the particle is $x = \cos(100\pi t)$

(a) The speed of the satellite is: $v_s = r\omega = 0.8 \times 100\pi = 80\pi$ m/s $= \underline{251 \text{ m/s}}$

(b) The speed of the particle at this point is:

$v = \left|\dfrac{dx}{dt}\right| = \left|0.8 \times 100\pi \times \sin(100\pi t)\right|_{t=0.005\,s} = \left|80\pi \times \sin\dfrac{\pi}{2}\right| = 80\pi$ m/s $= \underline{251 \text{ m/s,}}$

which is the same as the speed of the satellite.

15-7.   frequency $f = 80/\text{min} = 80 \text{ min}^{-1} \times \dfrac{1 \text{ min}}{60 \text{ s}} = 1.33$ Hz. $\omega = 2\pi f = 8.38$ rad/sec. The magnitude of

the force is

$F_{max} = ma_{max} = m\omega^2 A = (6.0 \text{ kg})(8.38 \text{ s}^{-1})^2(0.50 \text{ m}) = \underline{211 \text{ N}}.$

There is insufficient information given to find a velocity. The maximum speed is

$v_{max} = \omega A = 8.38 \text{ s}^{-1} \times 0.50 \text{ m} = \underline{4.2 \text{ m/s}}$

15-9. $x = A\cos(\omega t + \delta)$

When $t = 0$, $x = x_0 = A\cos\delta \Rightarrow \cos\delta = \dfrac{x_0}{A}$

$v = \dfrac{dx}{dt} = -A\omega\sin(\omega t + \delta)$

When $t = 0$, $v = v_0 = -A\omega\sin\delta \Rightarrow \sin\delta = -\dfrac{v_0}{A\omega}$

Because $\sin^2\delta + \cos^2\delta = 1$, $\left(-\dfrac{v_0}{A\omega}\right)^2 + \left(\dfrac{x_0}{A}\right)^2 = 1 \Rightarrow A^2 = x_0^2 + \dfrac{v_0^2}{\omega^2}$

$\Rightarrow A = \sqrt{x_0^2 + \dfrac{v_0^2}{\omega^2}},$

The phase angle can be expressed three different ways:

$\delta = \cos^{-1}\dfrac{x_0}{A} = \cos^{-1}\left(\dfrac{x_0}{\sqrt{x_0^2 + \dfrac{v_0^2}{\omega^2}}}\right)$ (from the equation for $x(0)$)

$\delta = -\sin^{-1}\dfrac{v_0}{\omega A} = -\sin^{-1}\left(\dfrac{v_0}{\omega\sqrt{x_0^2 + \dfrac{v_0^2}{\omega^2}}}\right) = -\sin^{-1}\left(\dfrac{v_0}{\sqrt{(\omega x_0)^2 + v_0^2}}\right)$ (from the equation for $v(0)$)

$\delta = -\tan^{-1}\left(\dfrac{v_0}{\omega x_0}\right)$ (by dividing $v(0)$ by $x(0)$)

15-11. $f = 250$ Hz. $\omega = 2\pi f = 2\pi \times 250$ rad/s $= 500\pi$ rad/s

$a_{max} = A\omega^2$ If $a_{max} = g = 9.81 \text{ m/s}^2$, the beads start to lift off.

$\Rightarrow A\omega^2 = 9.81\,\text{m/s}^2 \Rightarrow A = \dfrac{9.81 \text{ m/s}^2}{\omega^2} = \dfrac{9.81 \text{ m/s}^2}{(500\pi \text{ rad/s})^2} = \underline{3.98 \times 10^{-6} \text{ m}}.$

15-13. $x = A\cos(\omega t + \delta)$, $v = \dfrac{dx}{dt} = -A\omega\sin(\omega t + \delta)$

at $t = 0$, $x = 0 \Rightarrow 0 = \cos\delta \Rightarrow \delta$ can be $\pi/2$, $3\pi/2$.

at $t = 0$, $v > 0 \Rightarrow -A\omega\sin\delta > 0 \Rightarrow \delta = \underline{\dfrac{3\pi}{2}}$

15-17. Because $F = -kx$,

$$k = \left|\frac{F}{x}\right| = \frac{60 \text{ tons} \times 1000 \text{ kg/ton} \times 9.81 \text{ m/s}^2}{0.21 \text{ m}} = \underline{2.80 \times 10^6 \text{ N/m}}$$

angular frequency $\omega = \sqrt{\dfrac{k}{m}} = \sqrt{\dfrac{2.8 \times 10^6 \quad \text{N/m}}{7.10 \times 10^3 \text{kg}}} = \underline{19.9 \text{ rad/s}}$

frequency $f = \dfrac{\omega}{2\pi} = \dfrac{19.9 \quad \text{rad/s}}{2\pi \quad \text{rad}} = \underline{3.16 \text{ Hz}}$

15-23. $x = A\cos(\omega t + \delta), \quad v = \dfrac{dx}{dt} = -A\omega\sin(\omega t + \delta), \quad a = \dfrac{dv}{dt} = -A\omega^2\cos(\omega t + \delta)$

Because the mass is released from rest at a displacement of 20 cm from equilibrium, that will be the amplitude: $\underline{A = 0.20 \text{ m}}$.

$$\omega = \sqrt{\frac{k}{m}} = \sqrt{\frac{8.0 \quad \text{N/m}}{0.15 \quad \text{kg}}} = \underline{7.30 \quad \text{rad/s}}$$

At $t = 0$, $v = 0$, so $0 = -A\omega \cdot \sin\delta \Rightarrow \underline{\delta = 0}$

Therefore, $x = A\cos\omega t, \quad v = \dfrac{dx}{dt} = -A\omega\sin\omega t, \quad a = \dfrac{dv}{dt} = -A\omega^2\cos\omega t$

The maximum *speed* is
$v_{max} = A\omega = 0.20 \text{ m} \times 7.30 \text{ rad/s} = \underline{1.46 \text{ m/s}}$

The magnitude of the maximum acceleration is
$|a_{max}| = A\omega^2 = 0.20 \text{ m} \times (7.30 \text{ rad/s})^2 = \underline{10.7 \text{ m/s}^2}$

15-25. Angular frequency $\omega = \sqrt{\dfrac{k}{m}} = \sqrt{\dfrac{6.0 \times 10^5 \text{ N/m}}{0.5 \times 10^{-3} \text{ kg}}} = \underline{3.64 \times 10^4 \text{ rad/s}}$

frequency $f = \dfrac{\omega}{2\pi} = \dfrac{3.64 \times 10^4 \text{ rad/s}}{2\pi \text{ rad}} = \underline{5.51 \times 10^3 \text{ Hz}}$.

Because $f = \dfrac{\omega}{2\pi} = \dfrac{1}{2\pi}\sqrt{\dfrac{k}{m}}$, suppose frequency changes a small amount df, then df can be calculated by taking the differential of f:

$$df = -\frac{\sqrt{k}}{4\pi m^{3/2}} dm$$

The ratio $\dfrac{df}{f} = \dfrac{-\dfrac{\sqrt{k}}{4\pi m^{3/2}} dm}{\dfrac{1}{2\pi}\sqrt{\dfrac{k}{m}}} = -\dfrac{dm}{2m}$

Therefore, $\dfrac{dm}{m} = -2\dfrac{df}{f}$.

If $\dfrac{df}{f} = -0.01\%$, then $\dfrac{dm}{m} = -2 \times (-0.01\%) = 0.02\%$

$\Rightarrow$ The change of mass: $\Delta m = 0.02m = 0.02\% \times 0.5$ g $= \underline{1.0 \times 10^{-4}}$ g

Assume the thickness of the deposited film is t, then

$$t \times \text{Area} \times \text{density} = \Delta m \Rightarrow t = \frac{\Delta m}{\text{Area} \times \text{density}} = \frac{1 \times 10^{-4} \text{ g}}{2.0 \text{ cm}^2 \times 7.5 \text{ g/cm}^3} = \underline{6.7 \times 10^{-6}} \text{ cm}$$

15-35.   <u>If mass is added at the turning point, amplitude won't change.</u>

Because $f = \dfrac{\omega}{2\pi} = \dfrac{1}{2\pi}\sqrt{\dfrac{k}{m}}$,

if $m_{new} = 2\,m_{old}$, $f_{new} = \dfrac{1}{2\pi}\sqrt{\dfrac{k}{m_{new}}} = \dfrac{1}{2\pi\sqrt{2}}\sqrt{\dfrac{k}{m_{old}}} = \dfrac{1}{\sqrt{2}} f_{old} = \dfrac{1}{\sqrt{2}} \times 3.0$ Hz $= \underline{2.12 \text{ Hz}}$.

Because the amplitude does not change, $E = \dfrac{1}{2}kA^2$ won't   change.

Because $v_{max} = A\omega$, $v_{new}^{max} = A\omega_{new} = A(2\pi f_{new}) = 0.15$ m $\times 2\pi \times 2.12$ Hz $= \underline{2.0 \text{ m/s}}$.

Because $\left|a_{max}\right| = A\omega^2$, $\left|a_{new}^{max}\right| = A\omega_{new}^2 = A(2\pi f_{new})^2 = 0.15$ m $\times (2\pi \times 2.12$ Hz$)^2 = \underline{27 \text{ m/s}^2}$

15-37. $\omega = \dfrac{2\pi}{T} = \dfrac{2\pi}{1.5 \text{ sec}} = 4.19$ rad/sec

$\Rightarrow \quad \sqrt{\dfrac{k}{m}} = 4.19$ rad/sec $\Rightarrow \quad k = m\omega^2 = 0.5$ kg $\times (4.19)^2$ rad^2/sec$^2 = 8.78$ N/m

$\Rightarrow \quad E = \dfrac{1}{2}kA^2 = \dfrac{1}{2} \times 8.78\dfrac{\text{N}}{\text{m}} \bullet A^2 \Rightarrow \quad A = \sqrt{\dfrac{2E}{k}} = \sqrt{\dfrac{2 \times 0.5 \text{ J}}{8.78 \text{ N/m}}} = \underline{0.34 \text{ m}}$

15-43. $T = 2\pi\sqrt{\dfrac{l}{g}} = 2\pi\sqrt{\dfrac{27 \text{ m}}{9.81 \text{ m/s}^2}} = \underline{10.4 \text{ s}}$.

15-51. $T = 2\pi\sqrt{\dfrac{l}{g}} \Rightarrow l = \dfrac{T^2}{4\pi^2}g = \dfrac{10^2 \text{ s}^2}{4\pi^2} \times 9.81 \text{ m/s}^2 = \underline{24.8 \text{ m}}$

15-53. $T = 2\pi\sqrt{\dfrac{I}{mgd}}$, where d is the distance between the point of suspension and the center of mass. In

this case, $d = 0.9$ m.

$I = $ moment of inertia of the painting $= \dfrac{1}{2}mR^2$

$\Rightarrow T = 2\pi\sqrt{\dfrac{\dfrac{1}{2}mR^2}{mgd}} = 2\pi R\sqrt{\dfrac{1}{2gd}} = 2\pi \bullet 2 \text{ m} \bullet \dfrac{1}{\sqrt{2 \times 9.81 \text{ m/s}^2 \times 0.9 \text{ m}}} = \underline{3.0 \text{ s}}$

15-55.   (a) $\tau = I\alpha \Rightarrow -\kappa\theta = I\dfrac{d^2\theta}{dt^2}$,

where $I = $ moment of inertia of a disc $= \dfrac{1}{2}MR^2$

$\Rightarrow \dfrac{1}{2}MR^2\dfrac{d^2\theta}{dt^2} + \kappa\theta = 0 \Rightarrow \dfrac{d^2\theta}{dt^2} + \dfrac{2\kappa}{MR^2}\theta = 0$

The solution to this differential equation is $\theta = A\cos(\omega t + \delta)$,

where $\omega = \sqrt{\dfrac{2\kappa}{MR^2}}$

(b) The angular velocity is $\dot{\theta} = \dfrac{d\theta}{dt} = -A\omega\sin(\omega t + \delta)$. The maximum angular *speed* is

$$\left|\dot{\theta}_{max}\right| = A\omega = A\sqrt{\dfrac{2\kappa}{MR^2}} = \theta_0\sqrt{\dfrac{2\kappa}{MR^2}}$$

Note that $\dot{\theta}$ is not the same as ω!

(c) For $\omega = \dot{\theta}_{max}$, $\sqrt{\dfrac{2\kappa}{MR^2}} = \theta_0\sqrt{\dfrac{2\kappa}{MR^2}} \Rightarrow \theta_0 = 1$ radian

15-61. $T = 2\pi\sqrt{\dfrac{I}{mgd}}$

Because the stick swings around one of its ends, $I = \dfrac{1}{3}ml^2$. Therefore, $T = 2\pi\sqrt{\dfrac{\frac{1}{3}ml^2}{mgd}} = \dfrac{2\pi l}{\sqrt{3gd}}$.

The distance between the point of suspension and the center of mass is $d = 0.5$ m.
The length of the stick is $l = 1$ m.

Then $T = \dfrac{2\pi l}{\sqrt{3gd}} = \dfrac{2\pi \times 1\text{ m}}{\sqrt{3 \times 9.81\text{ m/s}^2 \times 0.5\text{ m}}} = 1.64$ s

15-77. As in Example 10, we'll use $Q = \dfrac{2\pi E}{\Delta E}$, where ΔE is the energy lost during

the first cycle and E is the pendulum's initial energy. To find this quantity,
use the hint that $\Delta E/E$ is the same for every cycle. If the energy decays by
ΔE to a value E_1 during the first cycle, then it will decay by ΔE_1 to E_2
during the second cycle with $\Delta E_1/E_1 = \Delta E/E$. So after one cycle the energy
would be $E_1 = E(1 - \Delta E/E)$ and after two cycles the energy would be

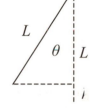

$E_2 = E_1(1 - \Delta E_1/E_1) = E(1 - \Delta E/E)^2$. After n cycles the energy would be

$E_n = E(1 - \Delta E/E)^n$, from which we get $\dfrac{\Delta E}{E} = 1 - \left(\dfrac{E_n}{E}\right)^{1/n}$.

When a pendulum of length L is raised through some angle θ, the mass at the end is lifted through
a height $L - L\cos\theta$. The potential energy of the mass is $mgh = mgL(1 - \cos\theta)$. If the pendulum is
released from this height, the equation gives the total energy of the pendulum as it swings.
According to the Math Help box following Eq. (15.46), the cosine function can be approximated
by $1 - \dfrac{\theta^2}{2}$ for small angles given in radians. Using this in the equation for the energy gives

$E = \dfrac{mgL\theta^2}{2}$. For a 1.5-m-long pendulum with an amplitude of 10° (0.175 radian), the initial

energy is $E = \dfrac{m(9.81\text{ m/s}^2)(1.5\text{ m})(0.175)^2}{2} = 0.224m$ J. (No mass is given in the problem, but it

will cancel in the calculation of Q.) When the amplitude is 4° (0.0698 radian), the final energy is

$E_{final} = 0.036m$ J. The period of the pendulum is $T = 2\pi\sqrt{\dfrac{L}{g}} = 2.46$ s, so 12 min represents

$\dfrac{12 \times 60}{2.46} \approx 293$ cycles. The energy loss per cycle is given by

$\dfrac{\Delta E}{E} = 1 - \left(\dfrac{0.036m}{0.224m}\right)^{1/293} = 6.22 \times 10^{-3}$. The final result is $Q = \dfrac{2\pi}{6.22 \times 10^{-3}} = 1.0 \times 10^3$.

15-79.   (a) Follow the method outlined in Problem 77. Here a mass of 25 kg is given. The initial and final amplitudes are 12° (0.209 radian) and 10° (0.175 radian), respectively.

$$E = \dfrac{mgL\theta_{init}^2}{2} = \dfrac{(25 \text{ kg})(9.81 \text{ m/s}^2)(3 \text{ m})(0.209)^2}{2} = 16.1 \text{ J}.$$

$$E_{final} = \dfrac{mgL\theta_{final}^2}{2} = \dfrac{(25 \text{ kg})(9.81 \text{ m/s}^2)(3 \text{ m})(0.175)^2}{2} = 11.3 \text{ J after five swings.}$$

$$\dfrac{\Delta E}{E} = 1 - \left(\dfrac{11.3}{16.1}\right)^{1/5} = 0.0684.$$

$$Q = \dfrac{2\pi E}{\Delta E} = \dfrac{2\pi}{0.0684} = \underline{91.9}$$

(b) The period is $T = 2\pi\sqrt{\dfrac{L}{g}} = 2\pi\sqrt{\dfrac{3 \text{ m}}{9.81 \text{ m/s}^2}} = 3.47$ s. The energy loss per cycle is

$|\Delta E| = \dfrac{2\pi E}{Q}$ from which we get the rate of energy loss:

$$\dfrac{|\Delta E|}{T} = \dfrac{2\pi E}{QT} = \dfrac{2\pi(16.1 \text{ J})}{(91.9)(3.47 \text{ s})} = \underline{0.317 \text{ W}}.$$

15-89.   (a) Since the first particle moves according to the equation $x = 0.30\cos(\dfrac{\pi}{4}t)$, it reaches its

midpoint when $x = 0 \Rightarrow \cos(\dfrac{\pi}{4}t) = 0 \Rightarrow \dfrac{\pi}{4}t = \dfrac{\pi}{2}, \dfrac{3\pi}{2}, \dfrac{5\pi}{2}, \cdots \Rightarrow t = \underline{2 \text{ s, 6 s, 10s, ...}}$. At the

turning points, $\dfrac{dx}{dt} = 0 \Rightarrow \sin(\dfrac{\pi}{4}t) = 0 \Rightarrow \dfrac{\pi}{4}t = 0, \pi, 2\pi, \cdots \Rightarrow t = \underline{0, 4s, 8s, ...}$

Turning Point	Midpoint	Turning Point	Turning Point	Midpoint	Turning Point
Satellite	Particle 1	Satellite	Particle 1	Satellite	Particle 1

120

(b) The second particle moves according to the equation $x' = 0.30\sin(\frac{\pi}{4}t)$. At its midpoint,

$$x' = 0 \Rightarrow \sin(\frac{\pi}{4}t) = 0 \Rightarrow \frac{\pi}{4}t = 0, \pi, 2\pi, \cdots \Rightarrow t = \underline{0, 4\text{ s, }8\text{s, }....}\text{ At its turning point,}$$

$$\frac{dx'}{dt} = 0 \Rightarrow \cos(\frac{\pi}{4}t) = 0 \Rightarrow \frac{\pi}{4}t = \frac{\pi}{2}, \frac{3\pi}{2}, \frac{5\pi}{2}, \cdots \Rightarrow t = \underline{2\text{ s, 6 s, 10s, }...}\;.$$

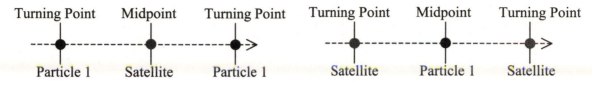

Turning Point   Midpoint   Turning Point   Turning Point   Midpoint   Turning Point

Particle 1   Satellite   Particle 1   Satellite   Particle 1   Satellite

(c) Find the time when $x = x'$:

$$0.3\cos(\frac{\pi}{4}t) = 0.3\sin(\frac{\pi}{4}t') = 0.3\cos(\frac{\pi}{4}t' - \frac{\pi}{2}) = 0.3\cos[\frac{\pi}{4}(t' - 2)]$$

We can see that when $t' = t+2$, we will have $x = x'$, i.e., if the first particle passes a certain point at time t, the second particle passes the same point at time $t+2$, or 2 s later.

15-93. $k_{new} = 2k \Rightarrow \omega_{new} = \sqrt{\frac{k_{new}}{m}} = \sqrt{2}\sqrt{\frac{k}{m}} = \sqrt{2}\omega_{old}$

$\Rightarrow f_{new} = \sqrt{2}f_{old} = \sqrt{2} \times 1.5\text{ Hz} = \underline{2.12\text{ Hz}}$

15-95. $f = \frac{1}{2\pi}\sqrt{\frac{k}{m}} = \frac{1}{2\pi}\sqrt{\frac{4.9 \times 10^3\text{ N/m}}{80\text{ kg}}} = \underline{1.25\text{ Hz}}$

Note that gravity does not affect the frequency of the motion. It only affects the equilibrium point.

15-97. $E = \frac{1}{2}kA^2 = \frac{1}{2} \times 6.0 \times 10^2\frac{\text{N}}{\text{m}} \times 0.25^2\text{ m}^2 = 18.75\text{ J}.$

At the equilibrium position, $U = 0$, so $K_{max} = E = 18.75\text{ J} \Rightarrow \frac{1}{2}mv_{max}^2 = 18.75\text{ J}$

$\Rightarrow v_{max} = \sqrt{\frac{2E}{m}} = \sqrt{\frac{2 \times 18.75\text{ J}}{3.0\text{ kg}}} = \underline{3.54\text{ m/s}}$

15-99. $T = 2\pi\sqrt{\frac{l}{g}} = 2\pi\sqrt{\frac{1.5\text{ m}}{9.8\text{ m/s}^2}} = \underline{2.46\text{s}}.$ To change to half the period,

$\frac{T}{2} = 2\pi\sqrt{\frac{l'}{g}} \Rightarrow \sqrt{\frac{l'}{g}} = \frac{1}{2}\sqrt{\frac{l}{g}} \Rightarrow l' = \frac{l}{4} = \underline{0.375\text{ m}}.$

CHAPTER 16 WAVES

16-3. From the graph, there is a negative peak at about 9 h GMT. There is a positive peak at about 15 h GMT. There are 12.5 cycles during this 6 h interval, so the frequency is about

$$f = \frac{12.5 \text{ cycles}}{6.0 \text{ h}} = 2.08 \text{ h}^{-1}. \quad \lambda = \frac{v}{f} = \frac{740 \text{ km/h}}{2.08 \text{ h}^{-1}} = \underline{356 \text{ km}}.$$

16-9. $\omega = 2\pi f = \dfrac{2\pi v}{\lambda} = \dfrac{(2\pi)(8.0 \text{ m/s})}{2.2 \text{ m}} = 22.8 \text{ s}^{-1}.$

(a) $v_{y,max} = \omega A = (22.8 \text{ s}^{-1})(0.012 \text{ m}) = \underline{0.27 \text{ m/s}}.$ The maximum transverse speed occurs when the particle goes through its equilibrium position (between the crests).

(b) $a_{y,max} = \omega^2 A = (22.8 \text{ s}^{-1})^2(0.012 \text{ m}) = \underline{6.2 \text{ m/s}^2}.$ The transverse acceleration has its maximum magnitude at the crests.

16-11. Using Eq. (16.11) and following the method presented in Example 3, $y = A\cos(kx_0 - \omega t)$. Taking $x_0 = 0$, $y = A\cos\omega t$. In this problem, $y = (0.020 \text{ m})\cos(9.0t)$, with t measured in s.

(a) Comparing the given equation with the standard, $\underline{A = 0.020 \text{ m}}$.

(b) $\omega = 9.0 \text{ s}^{-1} \Rightarrow f = \dfrac{\omega}{2\pi} = \underline{1.43 \text{ Hz}}.$

(c) $\lambda = \dfrac{v}{f} = \dfrac{14 \text{ m/s}}{1.43 \text{ Hz}} = \underline{9.8 \text{ m/s}}.$

16-13. (a) $T = 1/f = \lambda/v = (1.2 \text{ m})/(6.0 \text{ m/s}) = \underline{0.2 \text{ s}}.$ $f = 1/T = \underline{5.0 \text{ Hz}}.$ $\omega = 2\pi f = 10\pi \text{ s}^{-1}$ $= \underline{31.4 \text{ s}^{-1}}.$ $k = 2\pi/\lambda = (2\pi)/(1.2 \text{ m}) = \underline{5.2/\text{m}}$

(b) $y = A\cos(kx - \omega t + \delta)$, where the minus sign is chosen because the wave is propagating in the $+x$ direction. Since there is a crest at $x = 0$ at $t = 0$, $\delta = 0$. $A = 2.0 \text{ cm} = 0.020 \text{ m}$; $k = 5.2 \text{ m}^{-1}$; $\omega = 31.4 \text{ s}^{-1} \Rightarrow \underline{y = 0.020 \cos(5.2 x - 31.4 t)}$ where length is in meters, time in seconds.

16-15. $\lambda = \dfrac{v}{f}.$ Taking differentials gives $\Delta\lambda = \dfrac{\Delta v}{f}$ because f is constant. As the temperature changes from 30°C to 15°C, $\Delta v = v(15°\text{C}) - v(30°\text{C}) = 1440 \text{ m/s} - 1530 \text{ m/s} = -90 \text{ m/s}.$ Thus

$$\Delta\lambda = \frac{-90 \text{ m/s}}{440 \text{ Hz}} = -0.20 \text{ m}. \text{ The wavelength } \underline{\text{decreases by 20 cm}}.$$

16-17. $A = $ radius of the Ferris wheel $= 6.1 \text{ m}.$ $f = 6 \text{ rev/min} = 0.10 \text{ Hz}.$ The magnitude of the maximum acceleration is $|a_{max}| = 4\pi^2 f^2 A = (4\pi)(0.10 \text{ Hz})^2(6.1 \text{ m}) = \underline{2.41 \text{ m/s}^2}.$ From Problem 4,

$$\lambda = \frac{v}{f} = \frac{\sqrt{g\lambda/(2\pi)}}{f} \Rightarrow \lambda^2 = \frac{g\lambda}{2\pi f^2} \Rightarrow \lambda = \frac{g}{2\pi f^2} = \frac{9.81 \text{ m/s}^2}{(2\pi)(0.10 \text{ Hz})^2} = \underline{156 \text{ m}}.$$

16-19. Nine crests and troughs represent 9/2 complete waves, so $f = \dfrac{(9/2)\text{waves}}{15 \text{ s}} = \underline{0.30 \text{ Hz}}.$ The crest-to-trough distance is $\lambda/2$, so $\lambda = 2(0.75 \text{ m}) = \underline{1.5 \text{ m}}.$ $v = f\lambda = (0.30 \text{ Hz})(1.5 \text{ m}) = \underline{0.45 \text{ m/s}}.$

16-21. Density of copper $= \rho = 8.9$ g/cm$^3 = 8.9 \times 10^3$ kg/m^3. $\mu = \dfrac{M}{L} = \dfrac{\rho(\pi R^2 L)}{L} = \pi \rho R^2 =$

$\pi(8.9 \times 10^3 \text{ kg/m}^3)(5 \times 10^{-4} \text{ m}^2) = 6.99 \times 10^{-3}$ kg/m. $v = \sqrt{\dfrac{F}{\mu}} = \sqrt{\dfrac{75 \text{ N}}{6.99 \times 10^{-3} \text{ kg/m}}} = 104$ m/s.

The $\Delta t = \dfrac{L}{v} = \dfrac{100 \text{ m}}{104 \text{ m/s}} = \underline{0.97 \text{ s}}$.

16-29. The time difference is $\Delta t = \dfrac{d}{v_{20}} - \dfrac{d}{v_{100}}$, where d is the distance to the storm and v_{20}, v_{100} are the

speeds of the 20 m and 100 m waves respectively. Thus $d = \dfrac{\Delta t}{(1/v_{20}) - (1/v_{100})} =$

$\dfrac{(10 \text{ h})(3600 \text{ s/h})}{(2.8 \text{ m/s})^{-1} - (6.2 \text{ m/s})^{-1}} = 1.84 \times 10^5$ m. The storm is $\underline{184 \text{ km away}}$.

16-31. $v = \sqrt{\dfrac{F}{\mu}}$. The tension must be the same in both sections of the string,

so $v = \sqrt{\dfrac{F}{\mu}}$ and $v' = \sqrt{\dfrac{F}{\mu'}}$. Thus $\dfrac{v'}{v} = \sqrt{\dfrac{\mu}{\mu'}}$, or $\underline{v' = v\sqrt{\dfrac{\mu}{\mu'}}}$.

16-33. $F_1 = mg$, where $m = 30$ kg. At the point where the strings are connected together,

$2F_2 \cos 22.5° = F_1$, or $F_2 = \dfrac{F_1}{2 \cos 22.5°} = \dfrac{mg}{2 \cos 22.5°}$. Substituting numbers

gives $F_1 = 294$ N, $F_2 = 159$ N. The speed of a wave along each string is

$v_1 = \sqrt{\dfrac{F_1}{\mu}} = \sqrt{\dfrac{294 \text{ N}}{4.0 \times 10^{-3} \text{ kg/m}}} = 271$ m/s; $v_2 = \sqrt{\dfrac{159 \text{ N}}{4.0 \times 10^{-3} \text{ kg/m}}} = 200$ m/s.

The length of each string is L, so the total travel time for a pulse is

$\Delta t = \dfrac{L}{v_1} + \dfrac{L}{v_2} = (2 \text{ m})\left(\dfrac{1}{271 \text{ m/s}} + \dfrac{1}{200 \text{ m/s}}\right) = \underline{0.017 \text{ s}}$.

16-41. (a) $y_2 = 0.030 \sin(4.0x) = 0.030 \cos(4.0x - \pi/2)$. $\underline{\text{The sine function is } 90°, \text{ or } \pi/2 \text{ radians,}}$ $\underline{\text{behind the cosine function}}$.

(b) Follow the method described for finding the beat frequency in Eqs. (16.16 – 16.20).

$y = y_1 + y_2 = 0.030[\cos(4.0x) + \cos(4.0x - \pi/2)]$

$= 0.060\left\{\cos\left[\dfrac{4.0x - (4.0x - \pi/2)}{2}\right]\cos\left[\dfrac{4.0x + (4.0x - \pi/2)}{2}\right]\right\}$

$= 0.060\left(\cos\dfrac{\pi}{4}\right)\cos(4.0x - \pi/4)$

The final result is $\underline{y = 0.042 \cos\big((4.0x - \pi/4)\big)}$. The amplitude is $\underline{0.042 \text{ m}}$. The first maximum

(crest) in a cosine function occurs when the argument is zero, so the first crest occurs at

$x = \dfrac{\pi}{16}$ m $= \underline{0.20 \text{ m}}$.

16-43.   These waves are exactly 180° out of phase, so the total amplitude is the magnitude of the difference between the individual amplitudes: $A_{TOT} = |A_1 - A_2| = |A - 2A| = A$. Thus

$A_{TOT} = 6.0$ m. The wave number $k = 4.0$ m^{-1}, so $\lambda = \dfrac{2\pi}{4.0}$ m $= 1.57$ m. The angular frequency $\omega =$

5.0 s^{-1}, so $f = \dfrac{5.0}{2\pi}$ s$^{-1} = 0.796$ Hz.   At $x = 1.0$ m, $t = 1.0$ s,

$y = (6.0$ m$)[\cos(4.0 - 5.0) + \cos(4.0 - 5.0 - \pi) = \underline{-3.2 \text{ m}}$.

16-45.   Following the method in Eq. (16.19),

$$y_1 + y_2 = 2A\left\{\cos\left[\frac{(5.0x - 6.0t) - (6.0x - 7.0t)}{2}\right]\cos\left[\frac{(5.0x - 6.0t) + (6.0x - 7.0t)}{2}\right]\right\}$$

$$= 2A\left(\cos\frac{x-t}{2}\right)\cos\left(\frac{11x - 13t}{2}\right).$$

The first cosine term represents a long wavelength oscillation with wave number 0.5 m^{-1} and angular frequency 0.5 s^{-1}, and the second cosine term is a short wavelength oscillation with wave number 5.5 m^{-1} and angular frequency 6.5 s^{-1}. The short wavelength is $\lambda_{short} = \dfrac{2\pi}{5.5}$ m $= \underline{1.14 \text{ m}}$.

The long wavelength is $\lambda_{long} = \dfrac{2\pi}{0.5}$ m $= \underline{12.6 \text{ m}}$. The distance between zeros is $\lambda_{long}/2 = \underline{6.28 \text{ m}}$.

16-47.   (a) Using Eq. (16.16),

$$y_1 + y_2 = 2(0.030 \text{ m})\left\{\cos\left[\frac{(16x - 18t + 1.5) - (16x - 18t - 2.3)}{2}\right]\cos\left[\frac{(16x - 18t + 1.5) + (16x - 18t - 2.3)}{2}\right]\right\}$$

$$= (0.060 \text{ m})(\cos 1.9)\cos(6x - 18t - 0.4)$$

$$= -(0.019 \text{ m})\cos(6x - 18t - 0.4)$$

$\underline{A = 0.0194 \text{ m}}$.

(b) At $x = 0$, $t = 0$,

$y = -(0.0194 \text{ m})\cos(-0.4 \text{ radian}) = \underline{-0.0179 \text{ m}}$.

16-51. $f_{beat} = f_1 - f_2 = 4$ Hz. Therefore, the difference in the two is 4 Hz, so the frequency change must be 4 Hz.

$f = \sqrt{T/\omega}$ (λ const.) $\Rightarrow f\lambda = \sqrt{T/\mu}$

$f = \dfrac{\lambda}{\sqrt{\mu}}\sqrt{T}$

Therefore, $f_1/f_2 = \sqrt{T_1/T_2}$.

But $f_1/f_2 = 298/294$ (or 294/290) $= 1.014$ (both)

Thus, $(T_1/T_2) = (1.014)^2 = 1.028$

Therefore, tension needs to be increased (or decreased) 2.8%, but we don't know which (increase or decrease) is required.

16-59. $y = 2A\cos(\omega t)\cos(kx) = 10 \cos(3.0\pi t)\cos(\pi x)$. At $x = 0.25$ m,

$y = 10 \cos(3.0\pi t)\cos(0.25\pi) = 7.07\cos(3.0\pi t)$. The amplitude at this location is $\underline{7.07 \text{ m}}$.

$\left|v_{y,max}\right| = \omega A = (3\pi \text{ s}^{-1})(7.07 \text{ m}) = \underline{66.6 \text{ m/s}}$

$\left|a_{y,max}\right| = \omega v_{y,max} = (3\pi \text{ s}^{-1})(66.6 \text{ m/s}) = \underline{628 \text{ m/s}^2}$

16-61. If the plucked point is exactly at the center then the distance to each end is $L/2$ and the total distance traveled when the reflections return to the center is L. The time is

$$\Delta t = \frac{L}{v} = \frac{0.65 \text{ m}}{70 \text{ m/s}} = 9.3 \times 10^{-3} \text{ s, or } \underline{9.3 \text{ ms}}.$$ Since the ends of the string are fixed, the pulses are inverted upon reflection. If the string was pulled up when it was plucked, it will be displaced <u>down</u> when the reflections return to the center. The wavelength is $2(0.65 \text{ m}) = 1.30 \text{ m}$, so the vibration frequency is $f = \frac{v}{\lambda} = \frac{70 \text{ m/s}}{1.30 \text{ m}} = \underline{54 \text{ Hz}}.$

16-67. Let mirror be at $x = 0$. Incoming wave is $A \sin (kx - \omega t)$. Outgoing is $A \sin (kx - \omega t)$, so that $y = A \sin (kx - \omega t) + A \sin (kx + \omega t) = 2A \sin kx \cos \omega t$. The nearest antinode is when

$\sin kx = 1$ or $kx = \pi/2 \Rightarrow x = \pi/(2k) = (\pi/2)(\lambda/2\pi) = \lambda/4$

Nearest node at $\sin kx = 0$, or $kx = \pi \Rightarrow x = \pi/k = \lambda/2$.

$\lambda = 5.0 \times 10^{-7}$ m, so that the nearest antinode is at $\underline{1.25 \times 10^{-7} \text{ m}}$, and the nearest node is at $\underline{2.5 \times 10^{-7} \text{ m}}.$

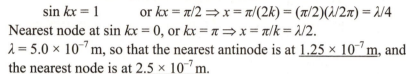

16-71. (a) Let θ be the angle of deflection from the normal position. Let T be the tension. Then $2T \sin \theta = 150$ N. Half the length of the rope is 4.5 m, and it deflected 0.070 m at its midpoint. Therefore

$$\sin \theta = (0.07/4.5). \quad T = \frac{150 \text{ N}}{2(0.070/4.5)} = \underline{4.82 \times 10^3 \text{ N}}.$$

(b) For the fundamental, $f_1 = \frac{1}{2L}\sqrt{\frac{T}{\mu}}$. Therefore, $f_1 = \frac{1}{2L}\sqrt{\frac{T}{\mu}} = \frac{1}{2(9.0 \text{ m})}\sqrt{\frac{4.82 \times 10^3 \text{ N}}{0.22 \text{ kg/m}}}$

$= \underline{8.2 \text{ Hz}}.$

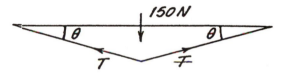

16-75. (a) $v(x) = \sqrt{\dfrac{F}{\mu(x)}} = \underline{\sqrt{\dfrac{F}{A + Bx}}}$

(b) $\lambda(x) = \dfrac{v(x)}{f} = \underline{\dfrac{1}{f}\sqrt{\dfrac{F}{A + Bx}}}$

(c) For each eigenfrequency, the wavelength must satisfy $\lambda(L) = \dfrac{2L}{n}$, $n = 1, 2, 3, \ldots$. Thus

$$\frac{2L}{n} = \frac{1}{f_n}\sqrt{\frac{F}{A + BL}} \Rightarrow \underline{f_n = \frac{n}{2L}\sqrt{\frac{F}{A + BL}}}.$$

16-77. $y = y_1 + y_2 = A_1 \cos(kx - \omega t) + A_2 \cos(kx + \omega t)$

$= A_1(\cos kx \cos \omega t + \sin kx \sin \omega t) + A_2(\cos kx \cos \omega t - \sin kx \sin \omega t)$

$= (A_1 + A_2)\cos kx \cos \omega t + (A_1 - A_2)\sin kx \sin \omega t$

Because $\sin \theta = \cos(\theta - \pi/2)$, this represents two standing waves that are $\pi/2$ or 90° out of phase with each other. The amplitude of the larger wave is $A_1 + A_2$ and these maxima occur where $\sin kx = 0$. The sine function is zero when the argument is $n\pi$, so the locations of the antinodes are $x_n = \dfrac{n\pi}{k}$.

The smaller standing wave has an amplitude of $|A_1 - A_2|$, and these occur where $\cos kx = 0$. The cosine function is zero when the argument is $\pi/2, 3\pi/2, 5\pi/2,\ldots$. The "nodes" (locations of minimum amplitude) are $x_n = \dfrac{(2n+1)\pi}{k}$, $n = 0, 1, 2,\ldots$.

16-81. Assuming the wave can be described by

$y = A \sin(kx - \omega t)$

$\dfrac{dy}{dt} = \omega a \cos(kx - \omega t)$

$\dfrac{d^2 y}{dt^2} = -\omega^2 A \sin(kx - \omega t)$

$v_{max} = A\omega$ and $a_{max} = A\omega^2$

where $A = 23.5/2$ m

$\omega = \dfrac{2\pi}{T} = \dfrac{2\pi}{15} = 0.419$ s

Therefore,

$v_{max} = \dfrac{23.5}{2} \times 0.419 = \underline{4.9 \text{ m/s}}$

$a_{max} = \dfrac{23.5}{2} \times (0.419)^2 = \underline{2.1 \text{ m/s}^2}$

16-85. The mass of the rope is $M = \pi\rho R^2 L$, where ρ is the density, R is the radius, and L is the length. The mass/length is $\mu = \dfrac{M}{L} = \pi\rho R^2 = \pi(1.1 \times 10^3 \text{ kg/m}^3)(4.6 \times 10^{-3}\text{ m})^2 = \underline{0.0731 \text{ kg/m}}$. The wave speed is $v = \sqrt{\dfrac{F}{\mu}} = \sqrt{\dfrac{5.0 \times 10^3 \text{ N}}{0.0731 \text{ kg/m}}} = \underline{261 \text{ m/s}}$.

16-87. (a) $\sum F_x = T_1 \cos 45° + T_2 \cos 30° - T = 0$

$\sum F_y = T_1 \sin 45° - T_2 \sin 30° = 0$

$\Rightarrow T_2 = T_1 \dfrac{\sin 45°}{\sin 30°} = T_1 \dfrac{1/\sqrt{2}}{1/2} = \sqrt{2}T_1.$

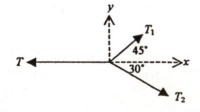

From the equation for the x components,

$\dfrac{T_1}{\sqrt{2}} + \sqrt{2}T_1\left(\dfrac{2}{\sqrt{3}}\right) - T = 0 \Rightarrow T_1\dfrac{1+\sqrt{3}}{\sqrt{2}} - T = 0 \Rightarrow T_1 = \underline{\dfrac{\sqrt{2}T}{1+\sqrt{3}}}.$

Then $T_2 = \underline{\dfrac{2T}{1+\sqrt{3}}}.$

(b) The wave speed is $v_n = \sqrt{\dfrac{T_n}{\mu}}$. All the strings are identical, so

$$\frac{v_1}{v} = \sqrt{\frac{T_1}{T}} = \sqrt{\frac{\sqrt{2}}{1+\sqrt{3}}} = 0.719 \Rightarrow \underline{v_1 = 7.2 \text{ m/s}}$$

$$\frac{v_2}{v} = \sqrt{\frac{T_2}{T}} = \sqrt{\frac{2}{1+\sqrt{3}}} = 0.856 \Rightarrow \underline{v_2 = 8.6 \text{ m/s}}$$

16-91. (a) The tension at any point along the rope is equal to the weight below that point. Thus at a distance x from the top of the rope, the tension will be $F(x) = \mu g(L - x)$, where μ is the mass per unit length of the rope. The speed is

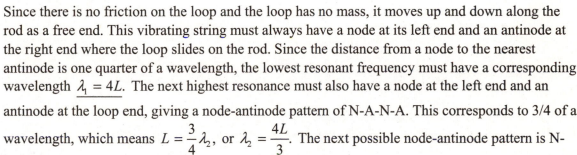

$$v(x) = \sqrt{\frac{F(x)}{\mu}} = \sqrt{g(L - x)}.$$

For $L = 20$ m, the speed at the top is $v(0) = \sqrt{(9.81 \text{ m/s}^2)(20 \text{ m})} = 14$ m/s. At the midpoint, $v(10 \text{ m}) = \sqrt{(9.81 \text{ m/s}^2)(20 \text{ m} - 10 \text{ m})} = 9.9$ m/s. At the bottom, $\underline{v(20 \text{ m}) = 0}$.

(b) $v(x) = \dfrac{dx}{dt} \Rightarrow dt = \dfrac{dx}{v(x)} = \dfrac{dx}{\sqrt{g(L - x)}}$. The time Δt for a pulse to travel from the top of the rope to the bottom is given by

$$\int_0^{\Delta t} dt = \int_0^L \frac{dx}{\sqrt{g(L-x)}} \Rightarrow \Delta t = -2\sqrt{\frac{L-x}{g}}\Bigg|_0^L = 2\sqrt{\frac{L}{g}}. \text{ The final result is}$$

$$\underline{\Delta t = 2\sqrt{\frac{20 \text{ m}}{9.81 \text{ m/s}^2}} = 2.9 \text{ s}.}$$

16-93. Since there is no friction on the loop and the loop has no mass, it moves up and down along the rod as a free end. This vibrating string must always have a node at its left end and an antinode at the right end where the loop slides on the rod. Since the distance from a node to the nearest antinode is one quarter of a wavelength, the lowest resonant frequency must have a corresponding wavelength $\lambda_1 = 4L$. The next highest resonance must also have a node at the left end and an antinode at the loop end, giving a node-antinode pattern of N-A-N-A. This corresponds to 3/4 of a wavelength, which means $L = \dfrac{3}{4}\lambda_2$, or $\underline{\lambda_2 = \dfrac{4L}{3}}$. The next possible node-antinode pattern is N-A-N-A-N-A, and L now corresponds to 5/4 of a wavelength. Thus $\underline{\lambda_3 = \dfrac{4L}{5}}$. The pattern of allowed wavelengths is given by $\underline{\lambda = 4L, \dfrac{4L}{3}, \dfrac{4L}{5}, \dfrac{4L}{7}, \dots}$.

127

17-9. $v_{max} = \sqrt{\dfrac{2I}{\rho_0 v}} = \sqrt{\dfrac{2(2 \times 10^4 \, \text{W/m}^2)}{(1.29 \, \text{kg/m}^3)(331 \, \text{m/s})}} = \underline{6.8 \, \text{m/s}}.$

This is an extremely loud sound, corresponding to 160 dB. Sounds of this intensity will cause structural damage to people and buildings.

17-11.   Since the tension is constant and $v = \sqrt{F/\mu}$, the speed of the wave on the string is constant. Since $v = f\lambda$, $\lambda = v/f$. Because the string is vibrating in its fundamental mode, $\lambda = 2L$, where L = length of string. Then $L = v/2f = (v/2)(1/f)$ or $L_1/L_0 = f_0/f_1$ where the subscript 0 indicates the frequency for the open string, 1 for the fretted string:

Tone	Frequency (Hz)	$L = \left(\dfrac{f_0}{f_1}\right) L_0$ (cm)	Spacing between frets (cm)
D	293.7	34.0	—
D#	311.2	32.1	1.9
E	329.7	30.3	1.8
F	349.2	28.6	1.7
F#	370.0	27.0	1.6
G	392.0	25.5	1.5
G#	415.3	24.0	1.5
A	440.0	22.7	1.3
A#	466.2	21.4	1.3
B	493.9	20.2	1.2
C	523.4	19.1	1.1
C#	554.4	18.0	1.1
D	587.4	17.0	1.0

17-13. $50 \, \text{dB} = 10 \log \left(\dfrac{I}{1.0 \times 10^{-12}} \right)$

$\log \left(\dfrac{I}{1.0 \times 10^{-12}} \right) = 5$

$I = 1.0 \times 10^{-12} \times 10^5 = \underline{1.0 \times 10^{-7} \, \text{w/m}^2}$

17-17.   Assuming each person produces the same intensity, $I(25 \, \text{people}) = I(50 \, \text{people})/2$. To find the change in [intensity], use $\Delta[\text{intensity level}](\text{dB}) = 10 \log \dfrac{I(25 \, \text{people})}{I_0} - 10 \log \dfrac{I(50 \, \text{people})}{I_0}$

$\Rightarrow \Delta[\text{intensity level}](\text{dB}) = 10 \log \dfrac{I(50 \, \text{people})}{2I_0} - 10 \log \dfrac{I(50 \, \text{people})}{I_0} = 10 \log \dfrac{1}{2}$

$\Rightarrow \Delta[\text{intensity level}](\text{dB}) = -3.0 \, \text{dB}.$

17-19. Let I_1 be the intensity of one machine and I_2 be the intensity of two machines. Assuming each machine produces the same intensity, $I_2 = 2I_1$. Then

[intensity level](dB) $= 10 \log \dfrac{I_2}{I_0} = 10 \log \dfrac{2I_1}{I_0} = 10 \log 2 + 10 \log \dfrac{I_1}{I_0}$

$\Rightarrow$ [intensity level](dB) $= 10 \log 2 + 80$ dB $= \underline{83\ \text{dB}}$

17-21. 70 dB $= 10 \log \left(\dfrac{I}{1.0 \times 10^{-12}} \right)$

$\log \left(\dfrac{I}{1.0 \times 10^{-12}} \right) = 7$

$I = 1.0 \times 10^{-12} \times 10^7 = \underline{1.0 \times 10^{-5}\ \text{w/m}^2}$

$I = \dfrac{P}{4\pi r^2}$

$\Rightarrow P = 4\pi r^2 I = 4\pi (30\ \text{m})^2 (1.0 \times 10^{-5}\ \text{W/m}^2) = \underline{0.11\ \text{W}}$

17-23. Intensity α 1/Area, Area α (diameter)^2.

Then $I_1 / I_0 = A_0 / A_1 = D_0^2 / D_1^2 = 8^2 / 0.7^2 = \underline{130\ \text{times more intense}}$

$\text{dB}_0 = 10 \log \left(\dfrac{I_0}{1.0 \times 10^{-12}} \right)$; $\text{dB}_1 = 10 \log \left(\dfrac{I_1}{1.0 \times 10^{-12}} \right) = 10 \log \left(\dfrac{130 I_0}{1.0 \times 10^{-12}} \right)$

Then, the increase in intensity level is

$\text{dB}_1 - \text{dB}_0 = 10 \log (I_1/I_0) = 10 \log 130 = \underline{21\ \text{dB}}$

17-27. Let d be the distance to the cliff. The time for the sound to travel from you to the cliff is d/v, and it takes the same time for the echo to return. The total time is $2t$, so

$d = \dfrac{vt}{2} = \dfrac{(331\ \text{m/s})(1.5\ \text{s})}{2} = \underline{249\ \text{m}}.$

17-33. From Table 17.3, $v = 344$ m/s at 20ºC. The length of the flute doesn't change, so

$f_{20} = \dfrac{v_{20}}{2L} = f_0 \dfrac{v_{20}}{v_0} = (261.7) \dfrac{344\ \text{m/s}}{331\ \text{m/s}} = \underline{272.0\ \text{Hz}}.$

To bring the flute back in tune, it must be lengthened. Taking differentials of the equation for f_0 gives

$df_0 = \dfrac{Ldv - vdL}{2L^2} = \left(\dfrac{v}{2L} \right) \left(\dfrac{dv}{v} - \dfrac{dL}{L} \right) \Rightarrow$

$\dfrac{df_0}{f_0} = \dfrac{dv}{v} - \dfrac{dL}{L}$

To bring the flute back in tune, we want $\dfrac{df_0}{f_0} = 0 \Rightarrow \dfrac{dv}{v} = \dfrac{dL}{L}$. Taking $v = 331$ m/s with $dv = 13$

m/s, we get $\dfrac{dL}{L} = \dfrac{dv}{v} = \dfrac{13}{331} = 0.039$. $\underline{\text{The flute must be lengthened by 3.9\%}}.$

17-39.   Assume the scuba divers are in sea water. Let t_1 = travel time through the pipe, t_2 = travel time through the water. Then

$$t_1 = \frac{d}{v_{pipe}}$$

$$t_2 = \frac{d}{v_{water}}$$

$$\Delta t = t_2 - t_1 = \frac{d}{v_{water}} - \frac{d}{v_{pipe}}$$

$$d = \frac{\Delta t}{\left(\dfrac{1}{v_{water}} - \dfrac{1}{v_{pipe}}\right)} = \frac{1.5 \text{ s}}{(1539 \text{ m/s})^{-1} - (5100 \text{ m/s})^{-1}} = \underline{3.3 \times 10^3 \text{m}\ \ (3.3 \text{ km})}$$

17-43.   Let d be the depth of the well. The time for the rock to fall is $t_1 = \sqrt{\dfrac{2d}{g}}$. The time for the sound to travel back is $t_2 = \dfrac{d}{v}$, where v is the speed of sound through air. The total time is

$t = t_1 + t_2 = \sqrt{\dfrac{2d}{g}} + \dfrac{d}{g}$. Move d/g to the other side, square both sides, and rearrange to get

$d^2 + 2\left(vt + \dfrac{v^2}{g}\right)d + (vt)^2 = 0$. Substitute numerical values ($t = 4.62$ s, $g = 9.81$ m/s^2, $v = 331$

m/s) to get $d^2 - 25395d + 2338514 = 0$. The solutions are

$$d = \frac{25395 \pm \sqrt{25395^2 - 4(2338514)}}{2} = 2.53 \times 10^4 \text{m, } 92.4 \text{ m}$$

The first answer is impossible because the sound could not travel that distance in less than 4.62 s. So the correct answer is $\underline{d = 92.4 \text{ m}}$.

17-47.   Let λ_n be the wavelength in the tube. Then since the ends of the tube must be pressure nodes, (displacement antinodes), we have

$$\left(\frac{\lambda_n}{2}\right)n = l \Rightarrow \lambda_n = \frac{2l}{n},$$

which gives

$$f_n = \frac{v}{\lambda_n} = n\frac{v}{2l}, n = 1, 2, 3$$

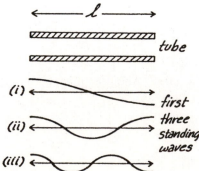

17-49. (a) In the fundamental mode $f = v/(2l)$, so that $l = v/(2f) = (331 \text{ m/s})/[2(261.7 \text{ Hz})] = \underline{0.632 \text{ m.}}$
(b) Calculate by using $l = v/(2f)$, where f is desired frequency

Tone	Frequency (Hz)	Length $= \dfrac{v}{2f}$ (m)	Spacing (m)
C	261.7	0.632	—
C#	277.2	0.597	0.035 = 3.5 cm
D	293.7	0.564	0.034 = 3.4 cm
D#	311.2	0.532	0.032 = 3.2 cm
E	329.7	0.502	0.030 = 3.0 cm
F	349.2	0.474	0.028 = 2.8 cm
F#	370.0	0.447	0.027 = 2.7 cm
G	392.0	0.422	0.025 = 2.5 cm
G#	415.3	0.399	0.024 = 2.4 cm
A	440.0	0.376	0.022 = 2.2 cm
A#	466.2	0.355	0.021 = 2.1 cm
B	493.9	0.335	0.020 = 2.0 cm
C	523.4	0.316	0.019 = 1.9 cm

17-51. Since the tube is closed at the ends, there will be a displacement node at both ends. If λ is the wavelength then, since there must be an integral number of half-wavelengths between ends,
$(\lambda_n/2)n = L \Rightarrow \lambda_n = 2L/n$
$\underline{\underline{f_n = v/\lambda_n = nv/(2L)}}$

17-57. The plane is moving with speed V_E at an altitude h. At some time it passes over the listener at L. Some time t later the shock wave, also traveling at speed V_E, passes L, and the plane has traveled a distance $V_E t$. From the geometry shown in the diagram, it is seen that the angle the shock wave makes with the ground is the same as the angle of the Mach cone. We have two equations giving the angle θ:

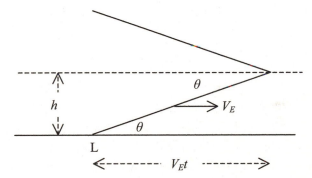

$$\tan \theta = \frac{h}{V_E t}$$

$$\sin \theta = \frac{v}{V_E}$$

where v is the speed of sound in air. Dividing the bottom equation by the top equation gives

$$\cos \theta = \frac{vt}{h} \Rightarrow \theta = \cos^{-1} \frac{vt}{h} = \cos^{-1} \frac{(331 \text{ m/s})(18 \text{ s})}{12000 \text{ m}} = 60.2°$$

Then

$$V_E = \frac{v}{\sin \theta} = \frac{331 \text{ m/s}}{\sin 60.2°} = \underline{381 \text{ m/s}}$$

17-59. The pulse arriving at the wall is Doppler shifted because the source (the bat) is approaching. The frequency of that pulse is $f_1 = f\left(\dfrac{1}{1-x}\right)$, where $x = v_{bat}/v$. This is the frequency of the pulse reflected back toward the bat. The frequency detected by the bat is Doppler-shifted a second time because the bat is now a receiver approaching the source. The detected frequency is

$f_2 = f_1(1+x) = f\dfrac{1+x}{1-x}$. Solving for x gives $x = \dfrac{f_2 - f}{f_2 + f} = \dfrac{800 \text{ Hz}}{50800 \text{ Hz} + 50000 \text{ Hz}} = 7.94 \times 10^{-3}$.

Thus $\underline{v_{bat} = 2.63 \text{ m/s}}$.

17-63. Let t_1 = the time the diver has been in free fall and t_2 = time for sound to travel back to the listener. At t_1 the diver is a distance $d = \dfrac{gt_1^2}{2}$ from the top of the cliff and is moving with speed $v_D = gt_1$. The relation between d and t_2 is $d = vt_2$, where v is the speed of sound. We want to know what pitch is heard 3.0 s after the diver jumps, so $t_1 + t_2 = 3 \Rightarrow t_2 = 3 - t_1 \Rightarrow d = v(3 - t_1)$. Setting the two expressions for d equal gives an equation for t_1:

$$3v - vt_1 = \dfrac{gt_1^2}{2}$$

or

$$gt_1^2 + 2vt_1 - 6v = 0$$

The solution is $t_1 = \dfrac{-2v + \sqrt{4v^2 + 24gv}}{2g} = \dfrac{-662 + \sqrt{662^2 + 24(9.81)(331)}}{2(9.81)} = 2.88$ s. (There is also a negative solution that has no physical meaning.) The speed of the diver at this instant is $v_D = (9.81 \text{ m/s}^2)(2.88 \text{ s}) = 28.2$ m/s. The diver is receding from the listeners at the top of the cliff, so the frequency they hear is $f' = \dfrac{f}{1 + v_D/v} = \dfrac{440 \text{ Hz}}{1 + 28.2/331} = \underline{405 \text{ Hz}}$.

17-65. In this problem, each driver hears two Doppler shifts: one caused by the motion of the source (the horn on the other car) and one caused by the motion of the receiver (the motion of each driver toward the other car). Car 1 has speed $v_1 = 90.0$ km/hr = 25.0 m/s, which is the emitter speed for the horn in car 1 and the receiver speed for the sound from the horn in car 2. Car 2 has speed $v_2 = 60.0$ km/hr = 16.7 m/s, which is the emitter speed for car 2 and the receiver speed for the sound from the horn in car 1. $f_{horn} = 524$ Hz for both cars.

(i)   To a stationary observer, horn from car 2 has frequency

$$f_2' = f_{horn}\left(\dfrac{1}{1 - v_2/v}\right)$$

This frequency is the source frequency for the driver in car 1, who hears

$$f_2'' = f_2'(1 + v_1/v) = f_{horn}\dfrac{1 + v_1/v}{1 - v_2/v} = (524 \text{ Hz})\dfrac{1 + (25.0/331)}{1 - (16.7/331)} = \underline{594 \text{ Hz}}$$

(ii) Similarly, the frequency heard by car 2 is Doppler-shifted twice because of the motion of the source toward the receiver and the motion of the receiver toward the source. Going through the same analysis, the frequency heard by the driver in car 2 is

$$f_1'' = f_{horn}\dfrac{1 + v_2/v}{1 - v_1/v} = (524 \text{ Hz})\dfrac{1 + (16.7/331)}{1 - (25.0/331)} = \underline{595 \text{ Hz}}$$

Note that here the source speed is v_1 and the receiver speed is v_2, the opposite of case (i).

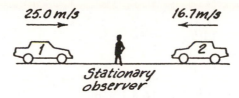

25.0 m/s   16.7 m/s

Stationary
observer

17-75.   The size depends on whether the object is located in air or tissue. Since this is a medical imager,
assume that the object is in tissue where $v \approx 1500$ m/s (the speed of sound in water). Then the
maximum size is close to half the wavelength:

$$d_{max} = \frac{\lambda}{2} = \frac{v}{2f} = \frac{1500 \text{ m/s}}{2(5.0 \times 10^6 \text{Hz})} = 1.5 \times 10^{-4} \text{m (0.15 mm)}$$

17-77. $$\lambda_{water} = \frac{v_{water}}{f} = \frac{1500 \text{ m/s}}{1.0 \times 10^6 \text{Hz}} = 1.5 \times 10^{-3} \text{m (1.5 mm)}$$

$$\lambda_{air} = \frac{v_{air}}{f} = \frac{331 \text{ m/s}}{1.0 \times 10^6 \text{Hz}} = 3.3 \times 10^{-4} \text{m (0.33 mm)}$$

Or use $\dfrac{v_{air}}{\lambda_{air}} = \dfrac{v_{water}}{\lambda_{water}} \Rightarrow \lambda_{water} = \lambda_{air} \dfrac{v_{water}}{v_{air}}$ to get the same answer.

17-79.   From the study of simple harmonic motion,

$$v_{max} = \omega A = 2\pi f A$$

$$|a_{max}| = \omega^2 A = 4\pi^2 f^2 A = \omega v_{max}$$

$$v_{max} = 2\pi(10 \times 10^6 \text{Hz})(8.0 \times 10^{-8} \text{m}) = 5.0 \text{ m/s}$$

$$|a_{max}| = 2\pi(10 \times 10^6 \text{Hz})(5.0 \text{ m/s}) = 3.2 \times 10^8 \text{ m/s}^2 \approx 3.2 \times 10^7 g$$

17-83. Δ[Intensity Level](dB) $= 10\log\dfrac{I_2}{I_0} - 10\log\dfrac{I_1}{I_0} = 10\log\dfrac{I_2}{I_1}$. $I_2 = 2I_1$, so

Δ[Intensity Level](dB) $= 10\log 2 = 3.0$ dB.

17-89.   (a) The man is at rest relative to the sound source, so he hears $f = 660$ Hz directly from it.
(b) The woman hears a tone Doppler shifted to a higher frequency because the source is
approaching her. Let $x = v_{boat}/v = 15/331 = 0.0453$. Then

$$f_{woman} = \frac{f}{1-x} = \frac{660 \text{ Hz}}{1 - 0.0453} = 691 \text{ Hz}.$$

(c) The echo heard by the man has been Doppler-shifted twice. The sound reaching the cliff has
been shifted because the boat is approaching the cliff, and the reflected sound has this increased
frequency. Then there is a second shift because the man, who is now the receiver, is approaching
the cliff. The speed of the source and the speed of the receiver are the same (the speed of the
boat), so the reflected frequency finally heard by the man is

$$f_{man} = \frac{f}{1-x} \times (1+x) = f\frac{1+x}{1-x} = (660 \text{ Hz})\frac{1 + 0.0453}{1 - 0.0453} = 723 \text{ Hz}.$$

CHAPTER 18   FLUID MECHANICS

18-1. Speed of flow $v = \dfrac{1}{A} \times$ rate of flow $= \dfrac{1}{A}\dfrac{\Delta V}{\Delta t}$

1 liter/min $= 1.67 \times 10^{-5}\, \text{m}^3/\text{s}$

(i) At hose: $A = \pi\left(\dfrac{3.81}{2} \times 10^{-2}\right)^2 = 1.14 \times 10^{-3}\, \text{m}^2$

Therefore, $v = \dfrac{95 \times 1.67 \times 10^{-5}}{1.14 \times 10^{-3}} = \underline{1.39\ \text{m/s}}$ At nozzle: $A = \pi R_2^{\,2}$

$v = \dfrac{95 \times 1.67 \times 10^{-5}}{A} = \dfrac{95 \times 1.67 \times 10^{-5}}{\pi\left(\dfrac{0.95}{2} \times 10^{-2}\right)^2} = \underline{22.3\ \text{m/s}}$

(ii) At hose $v = \dfrac{190 \times 1.67 \times 10^{-5}}{1.14 \times 10^{-3}} = \underline{2.8\ \text{m/s}}$

At nozzle $v = \dfrac{190 \times 1.67 \times 10^{5}}{\pi\left(\dfrac{1.27}{2} \times 10^{-2}\right)^2} = \underline{25.1\ \text{m/s}}$

(iii) At hose $v = \dfrac{284 \times 1.67 \times 10^{-5}}{1.14 \times 10^{-3}} = \underline{4.2\ \text{m/s}}$

At nozzle $v = \dfrac{284 \times 1.67 \times 10^{-5}}{\pi\left(\dfrac{1.59}{2} \times 10^{-2}\right)^2} = \underline{23.9\ \text{m/s}}$

18-3. Work to lift a mass m to a height h is $W = mgh$. The power is $P = \dfrac{dW}{dt} = gh\dfrac{dm}{dt}$. One liter of

water has a mass of 1 kg. Then

$P = (9.81\ \text{m/s}^2)(170\ \text{m})(26 \times 10^3\ \text{liters/min})(1\ \text{min}/60\ \text{s})(1\ \text{kg/liter}) = \underline{7.23 \times 10^5\ \text{W}}$

18-7. $\dfrac{dV}{dt} = \dfrac{0.50\ \text{liter}}{30\ \text{min}} \times \dfrac{1\ \text{min}}{60\ \text{s}} \times 10^{-3}\ \text{m}^3/\text{liter} = 2.78 \times 10^{-7}\ \text{m}^3/\text{s}$. In the tube,

$v_1 = \dfrac{dV/dt}{A_1} = \dfrac{2.78 \times 10^{-7}\ \text{m}^3/\text{s}}{\pi(1.0 \times 10^{-3}\,\text{m})^2} = 8.84 \times 10^{-2}\ \text{m/s, or } \underline{8.84\ \text{cm/s}}$. In the needle,

$v_2 = v_1\left(\dfrac{D_1}{D_2}\right)^2 = (8.84\ \text{cm/s})\left(\dfrac{2.0\ \text{mm}}{0.20\ \text{mm}}\right)^2 = 884\ \text{cm/s, or } \underline{8.84\ \text{m/s}}$.

18-15. Each tire supports $(1300/4\ \text{kg})(9.81\,\text{m/s}^2) = 3188\ \text{N}$. The normal force acting on each tire is
therefore 3188 N, which flattens the tire. If A is the area of contact, then $pA = 3188\ \text{N}$

$\Rightarrow A = \dfrac{3188\ \text{N}}{(2.4\ \text{atm})(1.01 \times 10^5\ \text{Pa/atm})} = 1.32 \times 10^{-2}\ \text{m}^2\text{, or } \underline{132\ \text{cm}^2}$.

18-21. The pressure on the outside of the suction cup is 1 atm, and the pressure on the inside is nearly
zero. The total force caused by the pressure difference must equal the weight of the boy, so

$pA = mg \Rightarrow A = \dfrac{mg}{p} = \dfrac{50\ \text{kg} \times 9.81\ \text{m/s}^2}{1.01 \times 10^5\ \text{Pa}} = \underline{4.86 \times 10^{-3}\,\text{m}^2}$ (or 48.6 cm^2).

18-25.   (a) Force $= A(\Delta p) = (0.06\ \text{m})^2 \times 1.01 \times 10^5\ \text{Pa} = \underline{360\ \text{N}}$.

(b) Since the sharpener is resting on a horizontal surface, the magnitude of the normal force is equal to the magnitude of the vertical force caused by the pressure difference. Thus maximum transverse force $= \mu_S N = 0.9 \times 360\ \text{N} = \underline{330\ \text{N}}$.

18-37.   From Eq. 14.20, $\left|\dfrac{\Delta V}{V}\right| = \dfrac{1}{B}p$, where we have used the definition of pressure given in this

chapter. The effect of atmospheric pressure is negligible at the depth in this problem. Thus
$\left|\dfrac{\Delta V}{V}\right| = \dfrac{4.54 \times 10^7\ \text{Pa}}{0.22 \times 10^{10}\ \text{Pa}} = 0.021,\ \text{or}\ \underline{2.1\%}$.

18-39.   The net force on the dam is caused by the overpressure from the water. The effect of atmospheric pressure is the same on both sides of the dam and causes no net force. For a strip of width dy located a distance y below the surface of the water, the pressure force caused by the water is $dF = \rho g y\, dA = 70\rho g y\, dy$. To find the total force, integrate over y from 0 to 30 m:

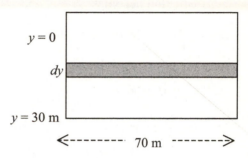

$F = \displaystyle\int_{y=0}^{y=30\ \text{m}} dF = 70\rho g \int_{0}^{30\ \text{m}} y\, dy = 35\rho g y^2 \Big|_{0}^{30\ \text{m}}$

$= (35)(1000\ \text{kg/m}^3)(9.81\ \text{m/s}^2)(30\ \text{m})^2 = \underline{3.1 \times 10^8\ \text{N}}$

18-47.   Mass of barrel $= 20\ \text{kg} + (0.12\text{m}^3)(730\ \text{kg/m}^3) = 108\ \text{kg}$
Volume of barrel $\approx 0.12\ \text{m}^3$
Therefore, the mass of water it displaces $= 0.12\ \text{m}^3 \times 1000\ \text{kg/m}^3 = 120\ \text{kg}$
Since the mass of the water displaced by the volume is greater than the mass of the barrel, the barrel floats.

18-49.   The buoyant force is equal to the weight of the air displaced. The total downward force is the weight of the helium in the balloon plus the weight of the balloon and the load hanging from the balloon. For the balloon to lift the load, the buoyant force must equal that total weight:

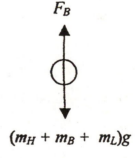

$F_B = \rho_{air} V g = \rho_H V g + m_B g + m_L g \Rightarrow V = \dfrac{m_B + m_L}{\rho_{air} - \rho_H}$

where V is the volume of the balloon. Then
$V = \dfrac{(4 + 150) \times 10^{-3}\ \text{kg}}{(1.29 - 0.18)\ \text{kg/m}^3} = 0.139\ \text{m}^3$

Assuming the balloon is spherical, its radius is
$R = \left(\dfrac{3V}{4\pi}\right)^{1/3} = \left[\dfrac{3(0.139\ \text{m}^3)}{4\pi}\right]^{1/3} = \underline{0.32\ \text{m}}$

18-53.   For him to submerge the ball, his weight must be at least equal to the weight of water displaced by the volume of the ball.
$mg = \rho_{water} V g = \dfrac{4\pi \rho_{water} R^3 g}{3} \Rightarrow m = \dfrac{4\pi \rho_{water} R^3}{3} = \dfrac{4\pi (1000\ \text{kg/m}^3)(0.30\ \text{m})^3}{3} = \underline{113\ \text{kg}}$ (The

friend must be very large!)

18-63. (a) Net force = buoyant force – weight if the positive direction is assumed to point up. Let x be the displacement from the surface of the water to the bottom of the boat. Then

$$\sum F = -A\rho g x - Mg = Ma = M\frac{d^2x}{dt^2}$$

$$\Rightarrow \frac{d^2x}{dt^2} = -\left(\frac{A\rho g}{M}x + g\right) = -\frac{A\rho g}{M}\left(x + \frac{M}{A\rho}\right)$$

Note the minus sign in front of the term with buoyant force term. This is required because x is measured down from the surface and the buoyant force points up. Let $x' = x + \dfrac{M}{A\rho}$. Then the equation becomes

$$\frac{d^2x'}{dt^2} = -\left(\frac{A\rho g}{M}\right)x', \text{ and as shown in Chapter 15, the motion of a particle with this equation is}$$

an oscillation with angular frequency

$$\omega = \sqrt{\frac{A\rho g}{M}}.$$

(b) Frequency $f = \omega/2\pi$

$$= \frac{1}{2\pi}\sqrt{\frac{(380 \times 60)\ \text{m}^2 \times (1025\ \text{kg/m}^3) \times 9.81\ \text{m/s}^2}{6.6 \times 10^8\ \text{kg}}} = \underline{0.094\ \text{Hz}}$$

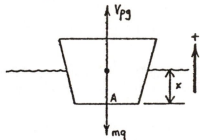

18-69. Bernoulli's equation says

$$\frac{1}{2}\rho v_{above}^2 + p_{above} = \frac{1}{2}\rho v_{below}^2 + p_{below}$$

where we assume the two points are at essentially the same level. Then

$$\Delta p = p_{below} - p_{above} = \frac{1}{2}\rho\left(v_{above}^2 - v_{below}^2\right) = \frac{1.29\ \text{kg/m}^3}{2}[(130\ \text{m/s})^2 - (115\ \text{m/s})^2] = 2.37 \times 10^3\ \text{Pa}$$

The net force on the wing is

$$F = (\Delta p)A = (2.37 \times 10^3\ \text{Pa})(8.0\ \text{m}^2) = \underline{1.9 \times 10^4\ \text{N}}$$

18-77. One way to solve this problem is to look at the physical significance of the terms in Bernoulli's equation. $\dfrac{1}{2}\rho v^2$ is the kinetic energy density (energy per unit volume) of the fluid; ρgh is the potential energy density; and p is the additional energy density produced by an external force doing work on the fluid. Bernoulli's equation expresses conservation of energy: in the absence of friction, energy density is constant throughout the fluid. In this problem, water is initially at rest under atmospheric pressure in the pond. The pump increases the energy density by lifting the water through a height h, increasing the pressure from 1 atm to 5.4 atm, and increasing the speed from essentially zero to the value that gives a volume flow of 1100 liters/min = 0.0183 m³/s through the 6.35 cm hose. Multiplying the increase in energy density by the volume flow will give the power required. Begin with the speed of water in the hose:

$$v = \frac{1}{A}\frac{dV}{dt} = \frac{0.0183 \text{ m}^3/\text{s}}{\pi\left(\frac{6.35}{2}\times 10^{-2}\text{ m}\right)^2} = 5.78 \text{ m/s}.$$

Use Bernoulli's equation to find the increase in energy density compared to the surface of the pond;

$$\Delta(E/V) = \frac{1}{2}\rho v^2 + \rho gh + \Delta p$$

$$= (500 \text{ kg/m}^3)(5.78 \text{ m/s})^2 + (1000 \text{ kg/m}^3)(9.81 \text{ m/s}^2)(4.5 \text{ m}) + (4.4 \text{ atm})(1.01\times 10^5 \text{ Pa/atm})$$

$$= 5.05\times 10^5 \text{ J/m}^3$$

The power required is

$$P = \Delta(E/V)\frac{dV}{dt} = (5.05\times 10^5 \text{ J/m}^3)(0.0183 \text{ m}^3/\text{s})(1 \text{ hp}/746 \text{ W}) = \underline{12.4 \text{ hp}}$$

18-81. If the speed of the plane relative to the air is v and the cross-sectional area of the window is A, then during time dt a tube of air with a volume of $Avdt$ strikes the window and essentially comes to rest. The force exerted on the window is equal to the rate of change in momentum of the air mass contained in this tube. 950 km/h = 250 m/s, so

$$F = \frac{dp}{dt} = \frac{(\rho_{air}Avdt)v}{dt} = \rho_{air}Av^2 = (1.29 \text{ kg/m}^3)(0.30 \text{ m})^2(250 \text{ m/s})^2 = \underline{7.3\times 10^3 \text{ N}}$$

18-89. When he is 2.0 m deep, the water pressure on his chest is

$$p_{water} = p_{atm} + \rho_{water}gh = 1.01\times 10^5 \text{ Pa} + 9.81 \text{ m/s}^2(1000 \text{ kg/m}^3)(2.0 \text{ m}) = \underline{1.21\times 10^5 \text{ Pa}}$$

The air pressure inside his chest is 1 atm $\underline{(1.01\times 10^5 \text{ Pa})}$, so the net pressure force on his chest is

$$F = (\Delta p)A = (0.20\times 10^5 \text{ Pa})(0.10 \text{ m}^2) = \underline{2.0\times 10^3 \text{ N}} \text{ (about 470 lb).}$$

18-91. (See Example 9.) The oil sinks until it displaces its own weight of water. Thus $\rho_{oil}V_{oil} = \rho_{water}V_{water}$. The volume of the oil above the surface of the water is

$$\Delta V = V_{oil} - V_{water} \Rightarrow \rho_{oil}V_{oil} = \rho_{water}(V_{oil} - \Delta V) \Rightarrow \frac{\Delta V}{V_{oil}} = \frac{\rho_{water} - \rho_{oil}}{\rho_{water}}. \text{ The oil and water both}$$

cover the same surface area A, so the total volume depends on the thickness of each volume. If the thickness of the oil is h_{oil} and the thickness of the displaced water is h_{water},

then $\Delta V = (h_{oil} - h_{water})A = (\Delta h)A$ and $\dfrac{\Delta h}{h_{oil}} = \dfrac{\rho_{water} - \rho_{oil}}{\rho_{water}}$

$$\Rightarrow \Delta h = \frac{1025 - 950}{1025}(10 \text{ cm}) = \underline{0.73 \text{ cm}}.$$

18-99. $\frac{1}{2}\rho v_1^2 + \rho gy_1 + p_1 = \frac{1}{2}\rho v_2^2 + \rho gy_2 + p_2$. Take point 1 to be at the top of the water in the tank, and point 2 is just below the hole in the bottom of the tank where the water runs out. The pressure at both points will be atmospheric pressure, so those terms cancel in the equation. Assume that the area A_1 of the water at the top is much larger than the area A_2 of the hole at the bottom, so

$$v_1 = \frac{A_2v_2}{A_1} \approx 0. \text{ Then } v_2 = \sqrt{2g(y_1-y_2)} = \sqrt{2(9.81 \text{ m/s}^2)(2 \text{ m})} = 6.26 \text{ m/s}.$$

The flow rate from the hole is $\dfrac{dV_2}{dt} = v_2A_2 = (6.26 \text{ m/s})(\pi)(0.01 \text{ m})^2 = \underline{2.0\times 10^{-3}\text{ m}^3/\text{s}}.$

19-3.   For iron, M_{Fe} = 56 g/mole.

$$-126.9°F = \frac{5}{9} \ (-126.9 - 32)°C = \underline{-88.3°C} = \underline{184.9°K}$$

The number of moles $n = \frac{m}{M_{Fe}} = \frac{0.50 \ g}{56 \ g/mole} = 8.9 \times 10^{-3}$ moles. So,

$$N = nN_A = (8.9 \times 10^{-3} \ moles)(6.02 \times 10^{23} \ molecules/mole) = \underline{5.3 \times 10^{21} \ atoms}$$

19-5.   From the given information, $\dfrac{m_{N_2}}{m_{N_2} + m_{O_2}} = 0.76$. With some algebraic manipulation this implies

$0.24 m_{N_2} = 0.76 m_{O_2}$ or $\dfrac{m_{N_2}}{m_{O_2}} = \dfrac{0.76}{0.24} = 3.2$ (1). In general, $m_{N_2} = M_{N_2} n_{N_2} = (28 \ g/mole) n_{N_2}$ and

$m_{O_2} = M_{O_2} n_{O_2} = (32 \ g/mole) n_{O_2}$. Substituting into (1) and solving,

$n_{N_2} = 3.2 \left(\dfrac{32}{28}\right) n_{O_2} = (3.6) n_{O_2}$ (2). Letting x = % O_2 by #, then by (2) $\dfrac{1-x}{x} = 3.6$. Solving for

$x = \dfrac{1}{4.6} = 0.22 = \underline{22\% \ O_2}$ and $1 - x = 1 - 0.22 = \underline{78\% \ N_2}$

19-7.   According to the Ideal-Gas Law, $\dfrac{p}{(n/V)} = RT$. Thus, the ratio of pressure to density is

proportional to the temperature. According to Eq. (17.5), $v \propto \sqrt{\dfrac{p_0}{\rho_0}} \propto \sqrt{T}$. If the temperature

drops from 293 K to 263 K, then the speed decreases by a factor $\sqrt{\dfrac{263}{293}} = 0.947$. Since the

frequency of a flute is directly proportional to the speed of sound, the frequency drops by a factor
of 0.947, from 261.7 Hz to 0.947 × 261.7 Hz = 247.9 Hz. This means that the frequency
decreases by 14 Hz.

19-11.   (a) Assume that the nitrogen and oxygen each behave like an ideal gas: $pV = nRT$, where

$n = m/M$. Therefore, $p = \dfrac{mRT}{MV}$. From the periodic table, M_{N_2} = 28 g/mole and

M_{O_2} = 32 g/mole.

For the oxygen tank, $p_{O_2} = \dfrac{(1.0 \ g)(8.31 \ J/mole \cdot K)(290 \ K)}{(32 \ g/mole)(1.0 \times 10^{-3} \ m^3)} = \underline{7.5 \times 10^4 \ Pa}$

For the nitrogen tank, $p_{N_2} = \dfrac{(1.0 \ g)(8.31 \ J/mole \cdot K)(290 \ K)}{(28 \ g/mole)(1.0 \times 10^{-3} \ m^3)} = \underline{8.6 \times 10^4 \ Pa}$

(b) When the two gases are combined into one tank at the same temperature, the total pressure
$p = p_{N_2} + p_{O_2} = (7.5 \times 10^4 \ Pa) + (8.6 \times 10^4 \ Pa) = \underline{1.6 \times 10^5 \ Pa}$

19-21. Assume that the air in the tire obeys the Ideal-Gas Law. The volume and number of molecules presumably stays constant throughout the day, so $\dfrac{p_i}{T_i} = \dfrac{p_f}{T_f} = $ constant. Specifically,

$$\frac{1.0 \text{ atm} + 4.0 \text{ atm}}{295 \text{ K}} = \frac{1.0 \text{ atm} + p_{f,\text{over}}}{311 \text{ K}}. \text{ Or, } p_{f,\text{over}} = \underline{4.3 \text{ atm}}$$

19-23. Using the Ideal-Gas Law, $n = \dfrac{m}{M}$, and assuming that the temperature is 273 K,

$$\frac{m}{V} = \frac{pM}{RT} = \frac{\left((400 \text{ atm})(1.01 \times 10^5 \text{ Pa/atm})\right)(0.028 \text{ kg/mole})}{(8.31 \text{ J/mole} \cdot \text{K})(273 \text{ K})} = \underline{500 \text{ kg/m}^3}$$

19-25. Using the Ideal-Gas Law and $n = \dfrac{m}{M}$,

$$\frac{m}{V} = \frac{pM}{RT} = \frac{\left((56 \text{ atm})(1.01 \times 10^5 \text{ Pa/atm})\right)(0.044 \text{ kg/mole})}{(8.31 \text{ J/mole} \cdot \text{K})(298 \text{ K})} = \underline{100 \text{ kg/m}^3}$$

19-27. The pressure 15 m below the lake is $p_{\text{atm}} + \rho g z$
$= 1.01 \times 10^5 \text{ N/m}^2 + (1000 \text{ kg/m}^3)(9.8 \text{ m/s}^2)(15 \text{ m})$
$= 2.48 \times 10^5 \text{ N/m}^2$
At the surface the pressure $= p_{\text{atm}} = 1.01 \times 10^5 \text{ N/m}^2$

Since $pV = $ const, $p_1 V_1 = p_2 V_2$, or $V_2 = V_1 \left(\dfrac{p_1}{p_2}\right)$.

But $V \alpha d^3$ where d is the diameter of the bubble. Thus,

$$d_2^3 = d_1^3 \left(\frac{p_1}{p_2}\right) \Rightarrow d_2 = d_1 \left(\frac{p_1}{p_2}\right)^{1/3}$$

$$d_2 = 1.0 \text{ cm} \left(\frac{2.48 \times 10^5}{1.01 \times 10^5}\right)^{1/3} = \underline{1.3 \text{ cm}}$$

19-29. Mass of CO_2 in the cylinder $= 8.2 - 5.9 = 2.3$ kg

Number of moles of $CO_2 = \dfrac{2.3 \text{kg}}{(12 + 2 \times 16)10^{-3} \text{kg/mol}} = 52.3$ mol

Therefore, $p = \dfrac{nRT}{V} = \dfrac{52.3 \times 8.31 \times 293}{2.8 \times 10^{-3}} = \underline{4.5 \times 10^7 \text{ N/m}^2}$

19-33. Number of moles of He contained in the balloon
$$n = pV / RT = \frac{3.2 \times 10^2 \times 8.5 \times 10^5}{8.31 \times 260} = 1.26 \times 10^5 \text{ mol.}$$

Net 'lift' force on the balloon = Buoyancy – Gravity $= n(0.029 - 0.004)g = $ payload $= mg$
Thus, $m = 1.26 \times 10^5 \times (0.025) = \underline{3150 \text{ kg}}$

Volume at STP, $V_1 = \dfrac{p}{p_1}\left(\dfrac{T_1}{T}\right) V = \dfrac{3.2 \times 10^2}{10^5}\left(\dfrac{273}{260}\right) 8.5 \times 10^5 = \underline{2.8 \times 10^3 \text{ m}^3}$

19-35. (a) Assume the temperature is constant. Then $p_1V_1 = p_2V_2$. But V (volume of air enclosed) $\propto h =$ height of air in the bell. Therefore, $p_1h_1 = p_2h_2$ or

$$h_2 = h_1\left(\frac{p_1}{p_2}\right), \text{ where } p_1 = p_{atm} = 1.01 \times 10^5 \text{ N/m}^2$$

$p_2 = p_{atm} + \rho gz$
$= 1.01 \times 10^5 \text{ N/m}^2 + (1000 \text{ kg/m}^3)(9.8 \text{ m/s}^2)(15 \text{ m})$
$= 2.48 \times 10^5 \text{ N/m}^2$

Therefore, $h_2 = 2m\left(\dfrac{1.01 \times 10^5}{2.48 \times 10^5}\right) = 0.81$ m. This means that the <u>water rises 1.2 m.</u>

(b) Pressure of air must equal the pressure of water at the lowest water level, i.e., at the bottom of the bell. The pressure there $p_3 = p_{atm} + \rho gZ$, where $Z = 15m + 1.2m$ of risen water (so we get to the bottom of the bell) $= 1.01 \times 10^5 \text{ N/m}^2 + (1000 \text{ kg/m}^3)(9.8 \text{ m/s}^2)(16.2 \text{ m}) = 2.6 \times 10^5 \text{ N/m}^2$
Original amount of air $n = p_1V/RT$ moles. Final amount of air $= (p_3V/RT)$ moles.
Thus, the number of moles of air added $= (p_3 - p_1)V/RT$ and the mass of air

$$\text{added} = \frac{(p_3 - p_1)V}{RT} M = \frac{(2.6 \times 10^5 - 1.01 \times 10^5)\, 3.53 \text{ m}^3}{(8.314)\, 288°K} \times 0.029 \text{ kg} = \underline{6.8 \text{ kg}}$$

19-39. Differentiating $p - p_0 = -\rho gy$ yields $dp = -\rho g\, dy$. Using the Ideal-Gas Law and $n = \dfrac{m}{M}$,

$$\rho = \frac{m}{V} = \frac{pM}{RT}. \text{ Substituting } \rho \text{ into the differential equation: } dp = -\frac{pMg}{RT}dy \text{ or, } \underline{\frac{dp}{p} = -\frac{Mg}{RT}dy}$$

19-41. Using $m = \dfrac{M}{N_A}$ and $v_{rms} = \sqrt{\dfrac{3kT}{m}}$,

$$v_{rms} = \sqrt{\frac{3kN_AT}{M}} = \sqrt{\frac{3(1.38 \times 10^{-23} \text{ J/K})(6.02 \times 10^{23} \text{ molecules/mole})(273 \text{ K})}{0.018 \text{ kg/mole}}} = \underline{615 \text{ m/s}}$$

19-43. Since $v_{rms} = \sqrt{\dfrac{3kT}{m}}$, then $\dfrac{v'_{rms}}{v_{rms}} = \sqrt{\dfrac{T'}{T}}$. That is, if $v'_{rms} = 2v_{rms}$, then

$$T' = 4T = 4 \times (300 \text{ K}) = \underline{1200 \text{ K}}$$

19-45. Using $m = \dfrac{M}{N_A}$ and $v_{rms} = \sqrt{\dfrac{3kT}{m}}$,

$$v_{rms} = \sqrt{\frac{3kN_AT}{M}} = \sqrt{\frac{3(1.38 \times 10^{-23} \text{ J/K})(6.02 \times 10^{23} \text{ molecules/mole})(673 \text{ K})}{1.0 \times 10^{-3} \text{ kg/mole}}} = \underline{4100 \text{ m/s}}$$

The average kinetic energy is $\dfrac{3}{2}kT = \dfrac{3}{2}(1.38 \times 10^{-23} \text{ J/K})(673 \text{ K}) = \underline{1.4 \times 10^{-20} \text{ J}}$

19-47. Using $m = \dfrac{M}{N_A}$ and $v_{rms} = \sqrt{\dfrac{3kT}{m}}$,

$$v_{rms} = \sqrt{\frac{3kN_AT}{M}} = \sqrt{\frac{3(1.38 \times 10^{-23} \text{ J/K})(6.02 \times 10^{23} \text{ molecules/mole})(5 \times 10^{-5} \text{ K})}{0.0855 \text{ kg/mole}}} = \underline{0.12 \text{ m/s}}$$

19-49. Average kinetic energy $= \frac{1}{2} m \overline{v^2}$. This can be expressed as $K = \frac{3}{2} kT$ (translational only).

Therefore, for *both* oxygen and nitrogen, $K = \frac{3}{2}(1.38 \times 10^{-23} \text{ J/}^\circ\text{K})(273^\circ\text{K}) = \underline{5.65 \times 10^{-21} \text{ J}}$

19-55. Solving $v_{rms} = \sqrt{\frac{3kT}{m}}$ for temperature results in

$$T = \frac{mv_{rms}^2}{3k} = \frac{\left(1.0 \times 10^{-12} \text{ kg}\right)\left(2.0 \times 10^{-2} \text{ m/s}\right)^2}{3\left(1.38 \times 10^{-23} \text{ J/K}\right)} = \underline{9.7 \times 10^6 \text{ K}}$$

19-57. Using $m = \frac{M}{N_A}$ and $v_{rms} = \sqrt{\frac{3kT}{m}}$,

$$v_{rms} = \sqrt{\frac{3kN_A T}{M}} = \sqrt{\frac{3\left(1.38 \times 10^{-23} \text{ J/K}\right)\left(6.02 \times 10^{23} \text{ molecules/mole}\right)\left(2.0 \times 10^{-4} \text{ K}\right)}{0.023 \text{ kg/mole}}} = \underline{0.47 \text{ m/s}}$$

19-59. Using the hint, the volume swept out per molecule with an effective radius R_0 going a distance l is cylindrically shaped with volume $V/N = \pi\left(2R_0\right)^2 l$. Solving for l yields the desired result,

$$l = \underline{\frac{1}{4\pi R_0^2 \left(N/V\right)}}$$

19-61. According to the given equation, the percent correction for the volume would be $\frac{-NV_m}{V}$.

Substituting into this expression the Ideal-Gas Law $V = \frac{nRT}{p}$ and $N = nN_A$ yields

$$\frac{-NV_m}{V} = \frac{-nN_A V_m}{\dfrac{nRT}{p}} = \frac{-pN_A V_m}{RT}.$$

(a) For 1.0 atm, $\dfrac{-pN_A V_m}{RT} = \dfrac{-\left(1.01 \times 10^5 \text{ Pa}\right)\left(6.02 \times 10^{23} \text{ molecules/mole}\right)\left(3.7 \times 10^{-29} \text{ m}^3\right)}{\left(8.31 \text{ J/mole} \cdot \text{K}\right)\left(298 \text{ K}\right)}$

$$= -9.1 \times 10^{-4} = \underline{-0.091\%}$$

(b) For 1000 atm,

$$\frac{-pN_A V_m}{RT} = \frac{-\left(\left(1000\right)\left(1.01 \times 10^5 \text{ Pa}\right)\right)\left(6.02 \times 10^{23} \text{ molecules/mole}\right)\left(3.7 \times 10^{-29} \text{ m}^3\right)}{\left(8.31 \text{ J/mole} \cdot \text{K}\right)\left(298 \text{ K}\right)}$$

$$= -0.91 = \underline{-91\%}$$

19-63. 1.0 kg of O_2 contains $\dfrac{1}{0.032 \text{ kg/mole}} = 31.25$ moles. For a diatomic ideal gas like oxygen the thermal kinetic energy

$$U = \frac{5}{2} NkT$$

$$= \left(\frac{5}{2}\right)\left[\left(31.25 \text{ moles}\right)\left(6.02 \times 10^{23} \text{ molecules/mole}\right)\right]\left(1.38 \times 10^{-23} \text{ J/K}\right)\left(293 \text{ K}\right)$$

$$= \underline{1.9 \times 10^5 \text{ J}}$$

In this case, $K_{trans} = \dfrac{3}{2} NkT$ and $K_{rot} = NkT$, so

Fraction translational $= \dfrac{3/2}{5/2} = \underline{0.6}$

Fraction rotational $= \underline{0.4}$

19-65. For each molecule, $K = \dfrac{3}{2} NkT$. For one mole at 300 K,

$$K = \left(\dfrac{3}{2}\right)(6.02 \times 10^{23} \text{ molecules})(1.38 \times 10^{-23} \text{ J/K})(300 \text{ K})$$

$$= \underline{3.7 \times 10^3 \text{ J}}$$

Increasing the temperature to 320 K, $K' = \left(\dfrac{T'}{T}\right)K = 1.07K$, a 7% increase in kinetic energy.

Increasing the pressure by 3 atm at constant temperature, $K' = K$, thus there is no change in the kinetic energy.

19-67. The number of moles of nitrogen can be calculated from the Ideal-Gas Law

$$n = \dfrac{pV}{RT} = \dfrac{\left((140 \text{ atm})(1.01 \times 10^5 \text{ Pa/atm})\right)(3.0 \times 10^{-2} \text{ m}^3)}{(8.31 \text{ J/mole} \cdot \text{K})(273 \text{ K})} = 187 \text{ moles}.$$ For a diatomic gas, the

change in internal energy $\Delta E = \dfrac{5}{2} nR\Delta T = \dfrac{5}{2}(187 \text{ moles})(8.31 \text{ J/mole} \cdot \text{K})(27 \text{ K}) = \underline{1.0 \times 10^5 \text{ J}}$

19-69. For seven total degrees of freedom, the total thermal energy is

$$\Delta E = \dfrac{7}{2} nR\Delta T = \dfrac{7}{2}(1.0 \text{ mole})(8.31 \text{ J/mole} \cdot \text{K})(10 \text{ K}) = \underline{291 \text{ J}}$$

19-71. One mole of CH_3OH has a mass of $(12 + 3 + 16 + 1) \text{ g} = \underline{32 \text{ g}}$

$$n = \dfrac{m}{M} = \dfrac{1.0 \times 10^3 \text{ g}}{32 \text{ g/mole}} = 31.25 \text{ mole, or}$$

$$N = nN_A = (31.25 \text{ mole})(6.02 \times 10^{23} \text{ molecules/mole}) = \underline{1.88 \times 10^{25} \text{ molecules}}$$

19-73. At each temperature, a pressure difference is responsible for the lift force. From the Ideal-Gas Law at the lower temperature $\Delta p = \dfrac{nRT}{V}$ and at the higher temperature $\Delta p' = \dfrac{nRT'}{V}$. From the definition of pressure, at the lower temperature the lift force $F = 1.2 \times 10^3 \text{ N} = \Delta pA$. At the higher temperature,

$$F' = \Delta p'A = \left(\dfrac{nRA}{V}\right)T' = (\Delta pA)\left(\dfrac{T'}{T}\right) = F\left(\dfrac{T'}{T}\right) = (1.2 \times 10^3 \text{ N})\left(\dfrac{308 \text{ K}}{278 \text{ K}}\right) = \underline{1.3 \times 10^3 \text{ N}}$$

19-75. For this problem use the average density of air $\rho = 1.29 \text{ kg/m}^3$.

$$\rho_{N_2} = 0.76\rho = (0.76)(1.29 \text{ kg/m}^3) = 0.98 \text{ kg/m}^3$$

$$\rho_{O_2} = 0.24\rho = (0.24)(1.29 \text{ kg/m}^3) = 0.31 \text{ kg/m}^3$$

$$m_{N_2} = \rho_{N_2}V = (0.98 \text{ kg/m}^3)(5.0 \times 10^{-3} \text{ m}^3) = 4.9 \times 10^{-3} \text{ kg}$$

$$m_{O_2} = \rho_{O_2}V = (0.31 \text{ kg/m}^3)(5.0 \times 10^{-3} \text{ m}^3) = 1.6 \times 10^{-3} \text{ kg}$$

For nitrogen $n = \dfrac{m}{M} = \dfrac{4.9 \times 10^{-3} \text{ kg}}{28 \times 10^{-3} \text{ kg/mole}} = 0.175 \text{ mole}$ or

$$N = nN_A = (0.175 \text{ mole})(6.02 \times 10^{23} \text{ molecules/mole}) = \underline{1.1 \times 10^{23} \text{ molecules}}$$

For oxygen $n = \dfrac{m}{M} = \dfrac{1.6 \times 10^{-3} \text{ kg}}{32 \times 10^{-3} \text{ kg/mole}} = 0.048 \text{ mole}$ or

$$N = nN_A = (0.048 \text{ mole})(6.02 \times 10^{23} \text{ molecules/mole}) = \underline{2.9 \times 10^{22} \text{ molecules}}$$

The total number of molecules is $(1.1 + 0.29) \times 10^{23} \text{ molecules} = \underline{1.4 \times 10^{23} \text{ molecules}}$

19-77.   (a) For the equal mixture of hydrogen ions and free electrons, the number of particles N in a

given mass m is $N = \dfrac{2m}{m_{H^+} + m_{e^-}}$, where m_{H^+} and m_{e^-} refer to the masses of individual

hydrogen ions and free electrons respectively and the factor of two is because there are two
particles for each pair of constituent particles. From this we can conclude that

$$\frac{N}{V} = \left(\frac{m}{V}\right)\frac{2}{m_{H^+} + m_{e^-}}$$

$$= (1.5 \times 10^5 \text{ kg/m}^3)\frac{2}{(1.67 \times 10^{-27} \text{ kg}) + (9.11 \times 10^{-31} \text{ kg})} = \underline{1.8 \times 10^{32} \text{ particles/m}^3}$$

Appling this to the Ideal-Gas Law,

$$p = \left(\frac{N}{V}\right)kT$$

$$= (1.8 \times 10^{32} \text{ particles/m}^3)(1.38 \times 10^{-23} \text{ J/K})(1.5 \times 10^7 \text{ K}) = \underline{3.7 \times 10^{16} \text{ Pa}}$$

(b) If the material in the center of the Sun were only hydrogen atoms, the number of particles N

in a given mass m would be $N = \dfrac{m}{m_H}$. From this we can conclude that

$$\frac{N}{V} = \left(\frac{m}{V}\right)\frac{1}{m_H} = (1.5 \times 10^5 \text{ kg/m}^3)\frac{1}{(1.67 \times 10^{-27} \text{ kg})} = \underline{9.0 \times 10^{31} \text{ particles/m}^3}$$

Appling this to the Ideal-Gas Law,

$$p = \left(\frac{N}{V}\right)kT$$

$$= (9.0 \times 10^{31} \text{ particles/m}^3)(1.38 \times 10^{-23} \text{ J/K})(1.5 \times 10^7 \text{ K}) = \underline{1.9 \times 10^{16} \text{ Pa}}$$

(c) If the material in the center of the Sun were only diatomic hydrogen molecules, the number of

particles N in a given mass m would be $N = \dfrac{m}{m_{H_2}}$. From this we can conclude that

$$\frac{N}{V} = \left(\frac{m}{V}\right)\frac{1}{m_{H_2}} = (1.5 \times 10^5 \text{ kg/m}^3)\frac{1}{2(1.67 \times 10^{-27} \text{ kg})} = \underline{4.5 \times 10^{31} \text{ particles/m}^3}$$

Applying this to the Ideal-Gas Law,

$$p = \left(\frac{N}{V}\right)kT$$

$$= (4.5 \times 10^{31} \text{ particles/m}^3)(1.38 \times 10^{-23} \text{ J/K})(1.5 \times 10^7 \text{ K}) = \underline{9.3 \times 10^{15} \text{ Pa}}$$

19-79. In general from the Ideal-Gas Law at constant temperature $pV = p'V'$. So,

$$\frac{\Delta V}{V} = \frac{V - V'}{V} = 1 - \frac{V'}{V} = 1 - \frac{p}{p'} = \frac{p' - p}{p'} = \frac{\Delta p}{p'}. \text{ This can be rearranged to } V = \Delta V\left(\frac{p'}{\Delta p}\right),$$

where ΔV is the *decrease* in volume and Δp is the corresponding *increase* in pressure.

Furthermore, $\frac{p'}{\Delta p} = \frac{p + \Delta p}{\Delta p} = 1 + \frac{p}{\Delta p}$ and $V = \Delta V\left(1 + \frac{p}{\Delta p}\right) \approx \Delta V\left(\frac{p}{\Delta p}\right)$ for $\Delta p \ll p$. Using the

specifics of the problem, $V_c = V_s + V$ or $V_s = V_c - V = V_c - \Delta V\left(\frac{p}{\Delta p}\right)$

19-81. Using $m = \dfrac{M}{N_A}$ and $v_{rms} = \sqrt{\dfrac{3kT}{m}}$,

$$v_{rms} = \sqrt{\frac{3kN_A T}{M}} = \sqrt{\frac{3\left(1.38 \times 10^{-23} \text{ J/K}\right)\left(6.02 \times 10^{23} \text{ molecules/mole}\right)\left(273 \text{ K}\right)}{0.018 \text{ kg/mole}}} = \underline{615 \text{ m/s}}$$

19-83. Since $v_{rms} = \sqrt{\dfrac{3kT}{m}}$, then $\dfrac{v'_{rms}}{v_{rms}} = \sqrt{\dfrac{T'}{T}}$.

(a) $\dfrac{v'_{rms}}{v_{rms}} = \sqrt{\dfrac{2T}{T}} = \underline{\sqrt{2}}$

(b) $\dfrac{v'_{rms}}{v_{rms}} = \sqrt{\dfrac{T}{T}} = \underline{1}$

(c) $\dfrac{v'_{rms}}{v_{rms}} = \sqrt{\dfrac{T}{T}} = \underline{1}$

(d) $\dfrac{v'_{rms}}{v_{rms}} = \sqrt{\dfrac{T/4}{T}} = \underline{1/2}$

19-85. In order for $K_i = K_f$, then $K_{H_2} + K_{O_2} = K_{H_2O}$. For the hydrogen and the oxygen (both diatomic),

$K = \dfrac{5}{2}nRT$. For the water (nonlinear polyatomic), $K = 3nRT$. Substituting into the kinetic

energy conservation equation:

$\dfrac{5}{2}(3 \text{ moles})(8.31 \text{ J/mole} \cdot \text{K})(300 \text{ K}) = 3(2 \text{ moles})(8.31 \text{ J/mole} \cdot \text{K})(T_f)$, where '3 moles' was

used on the left side of the equation because there were two moles of hydrogen and one mole of
oxygen. Solving for $T_f = \underline{375 \text{ K}}$

CHAPTER 20   HEAT

20-1.   Since no heat is lost to the surroundings, the temperature of the water will increase as heat Q is added. The amount of temperature increase ΔT is determined by the mass of the water m and its specific heat c. The heater converts electricity at a rate $P = 620$ W to increase the temperature of the water. This power P is equal to a rate of 620 J/s. To determine the time required to increase the temperature of the water, we recognize that

$$P = \frac{Q}{t} = \frac{mc\Delta T}{t}$$

$$t = \frac{mc\Delta T}{P} = \frac{(1.0 \text{ liter})\left(\frac{1.0 \text{ kg}}{1.0 \text{ liter}}\right)\left(4187 \text{ J/(kg}\cdot{}^\circ\text{C})\right)(80\ ^\circ\text{C})}{620 \text{ J/s}} = \underline{540 \text{ s}}$$

20-3.   Assume that all of the gravitational potential energy (mgh) of the water is converted into heat Q. Then, the temperature change of the water can be determined by rearranging the equation $Q = mc\Delta T$.

$$\Delta T = \frac{Q}{mc} = \frac{mgh}{mc} = \frac{gh}{c} = \frac{(9.81 \text{ m/s}^2)(120 \text{ m})}{4187 \text{ J/(kg}\cdot{}^\circ\text{C})} = \underline{0.28\ ^\circ\text{C}}$$

Note that the mass of the water drops out of the equation. Thus, this would be the temperature change regardless of the actual amount of water that fell.

20-7.   The power P you exert rotates the paddles, which, in turn, do mechanical work on the water. The process converts chemical energy in your body into mechanical work and then into a temperature increase of the water.

$$P = \frac{Q}{t} = \frac{mc\Delta T}{t}$$

$$t = \frac{mc\Delta T}{P} = \frac{4.0 \text{ liters}\left(\frac{1.0 \text{ kg}}{1.0 \text{ liter}}\right)\left(4186 \text{ J/(kg}\cdot{}^\circ\text{C})\right)(5.0\ ^\circ\text{C})}{0.15 \text{ hp}\left(\frac{745.7 \text{ W}}{1 \text{ hp}}\right)\left(\frac{1 \text{ J/s}}{1 \text{ W}}\right)} = \underline{750 \text{ s}}$$

20-9.   Assume that all of the energy contained in the snack is converted by the body into gravitational potential energy at a height H above the starting location.
$Q = mgH = mgnh$

$$n = \frac{Q}{mgh} = \frac{350 \text{ kcal}\left(\frac{4187 \text{ J}}{1 \text{ kcal}}\right)}{(70 \text{ kg})(9.81 \text{ m/s}^2)(0.25 \text{ m})} = \underline{8500 \text{ steps}}$$

In reality, the body is not completely efficient; and it also uses energy for other purposes, so one would have completely used up the energy in the snack in reaching a much lower altitude than indicated.

20-13.   Friction processes generally cause a portion of the mechanical energy of a system to be "lost" in heating the components that are rubbing. Without a means of cooling, this heating could damage components. Therefore, it is necessary to cool the components, in this case, with cooling water. The water passes over the parts and extracts heat at a rate P from them. As a result, the temperature of the water changes by

$$P = \frac{Q}{t} = \left(\frac{m}{t}\right)c\Delta T$$

$$\Delta T = \frac{P}{(m/t)c} = \frac{(300 \text{ W})}{\left(2.5\ \frac{\text{liters}}{\text{min}}\right)\left(\frac{1.0 \text{ kg}}{1 \text{ liter}}\right)\left(\frac{1 \text{ min}}{60 \text{ s}}\right)\left(4187 \text{ J/(kg}\cdot{}^\circ\text{C})\right)} = \underline{1.7\ ^\circ\text{C}}$$

20-15. Since the kinetic energy of the water remains constant, the increase in kinetic energy that would normally accompany a decrease in potential energy is being dissipated by friction. The rate of energy loss per kilometer is

$$P = \frac{mg\Delta h}{t} = \frac{\rho V g \Delta h}{t} = \rho g \Delta h \left(\frac{V}{t} \right)$$

The mass of the water is the product of its density ρ and its volume V. This is also equal to the rate at which heat is added to the water per kilometer.

$$\rho g \Delta h \left(\frac{V}{t} \right) = \frac{mc\Delta T}{t} = \left(\frac{m}{t} \right) c \Delta T$$

$$\Delta T = \frac{\rho g \Delta h (V/t)}{(m/t)c} = \frac{g \Delta h}{c} = \frac{(9.81 \text{ m/s}^2)(0.074 \text{ m})}{4187 \text{ J/(kg} \cdot \text{°C)}} = \underline{1.7 \times 10^{-4} \text{°C/km}}$$

20-17. The power used is 2.24×10^{10} W. The total incident solar power is

$$P_S = (1000 \text{ W/m}^2)(850 \text{ km}^2)\left(\frac{1000 \text{ m}}{1 \text{ km}}\right)^2 = 8.5 \times 10^{11} \text{ W}$$

$$\frac{2.24 \times 10^{10} \text{ W}}{8.5 \times 10^{11} \text{ W}} = \underline{2.6\%}$$

Therefore, the power used is 2.6% of the incident solar power. This is enough to slightly increase the local temperature.

20-21. (a) Consider a small amount of water dm approaching the bottom of the water wheel with a speed v_1. The angular momentum of dm with respect to the axis of rotation of the wheel is $(dm)v_1R$. As it leaves the wheel, its angular momentum has changed to $(dm)v_2R$. By the principle of the conservation of angular momentum, the angular momentum lost by the water must be gained by the water wheel, assuming frictional losses are negligible. The rate at which the wheel gains angular momentum from the water is

$$\tau = \frac{dL}{dt} = \frac{dm}{dt}(v_1 - v_2)R$$

$$= (300 \text{ kg/s})(5.0 \text{ m/s} - 2.5 \text{ m/s})(2.2 \text{ m}) = 1650 \text{ N} \cdot \text{m or } \underline{1.7 \times 10^3 \text{ N} \cdot \text{m}}$$

(b) The power delivered to the wheel is the product of the torque and the angular speed.

$$P = \tau \omega = (1650 \text{ N} \cdot \text{m})(1.4 \text{ rad/s}) = \underline{2.3 \times 10^3 \text{ W}}$$

(c) Each second, 300 kg of water pass through the wheel and lose kinetic energy in the process. Some of this kinetic energy becomes the rotational kinetic energy of the wheel and some goes into heating the water. The rate of energy lost by 300 kg of water each second is

$$\frac{\Delta K}{\Delta t} = \frac{\frac{1}{2}m(v_2 - v_1)^2}{\Delta t}$$

$$= \frac{\frac{1}{2}(300 \text{ kg})\left[(2.5 \text{ m/s})^2 - (5.0 \text{ m/s})^2\right]}{1 \text{ s}} = 2813 \text{ J/s}$$

The difference in the power delivered to the wheel and the power lost by the water is the power that goes into heating the water.

$$\Delta P = \frac{Q}{t} = \left(\frac{m}{t}\right)c\Delta T$$

$$\Delta T = \frac{\Delta P}{(m/t)c} = \frac{(2813\ \text{W} - 2310\ \text{W})}{(300\ \text{kg/s})(4187\ \text{J/(kg} \cdot {}^\circ\text{C}))} = \underline{4.0 \times 10^{-4}\,{}^\circ\text{C}}$$

20-23. The Eiffel Tower was built of 7300 tons of iron, which has a coefficient of linear thermal expansion of $12 \times 10^{-6}/{}^\circ\text{C}$. As the temperature increases, the dimensions of the tower increase. The temperature increase required to increase the height by 0.10 m is found using
$\Delta L = \alpha L \Delta T$

$$\Delta T = \frac{\Delta L}{\alpha L} = \frac{0.10\ \text{m}}{(12 \times 10^{-6}/{}^\circ\text{C})(312\ \text{m})} = \underline{27{}^\circ\text{C}}$$

20-25. When the copper pipe is heated, its radius will increase linearly with temperature, in similar manner as the length of a heated rod will lengthen. In this case, the diameter of the sleeve must increase by 0.002 cm to just barely fit over the steel rod. The temperature increase required for this thermal expansion can be found from
$\Delta L = \alpha L \Delta T$

$$\Delta T = \frac{\Delta L}{\alpha L} = \frac{0.002\ \text{cm}}{(17 \times 10^{-6}/{}^\circ\text{C})(0.998\ \text{cm})} = 118{}^\circ\text{C}$$

Therefore, the sleeve must be heated an additional 118°C from 18°C to $\underline{136{}^\circ\text{C}}$.

20-29. When the plastic frame is immersed in the bath, it will linearly expand in each dimension. Thus, the openings for the lenses will also expand. In this case, we want the opening to be 0.75% larger than its original size. This is the same as indicating that $\Delta L/L$ is equal to 0.0075.
$\Delta L = \alpha L \Delta T$

$$\Delta T = \frac{1}{\alpha}\frac{\Delta L}{L} = \frac{1}{2.0 \times 10^{-4}/{}^\circ\text{C}}(0.0075) = \underline{38{}^\circ\text{C}}$$

20-31. The volume of the alcohol will expand in all three dimensions as the system is heated. At the same time, the volume of the glass container also increases, but not as much since the thermal expansion coefficient is smaller for glass. To determine how much alcohol spills, we must determine the difference in the volume increase of alcohol to that of the container.
$$\Delta V_{\text{alcohol}} - \Delta V_{\text{glass}} = (\beta_{\text{alcohol}} - 3\alpha_{\text{glass}})V\Delta T$$
$$= \left[(1.0 \times 10^{-3}/{}^\circ\text{C}) - 3(9.0 \times 10^{-6}/{}^\circ\text{C})\right](1.0\ \text{liter})(75{}^\circ\text{C} - (-110{}^\circ\text{C}))$$
$$= \underline{0.18\ \text{liter}}$$

20-33. The work done in expansion is $W = \rho\Delta V$, where the volume change is $\Delta V = \beta V\Delta T$.
$$W = P\Delta V = P\beta V\Delta T = P3\alpha V\Delta T$$
$$= (1.0\ \text{atm})\left(\tfrac{1.013 \times 10^5\ \text{N/m}^2}{1.0\ \text{atm}}\right)(3)(12 \times 10^{-6}/{}^\circ\text{C})(1.0\ \text{kg})\left(\tfrac{1\ \text{m}^3}{7900\ \text{kg}}\right)(60{}^\circ\text{C}) = \underline{0.028\ \text{J}}$$

The heat added to the cube is
$$Q = mc\Delta T = (1.0\ \text{kg})(445\ \text{J/(kg} \cdot {}^\circ\text{C}))(60{}^\circ\text{C}) = \underline{2.7 \times 10^4\ \text{J}}$$

The ratio of the heat added to the cube and the work done is
$$\frac{Q}{W} = \frac{0.028\ \text{J}}{2.7 \times 10^4\ \text{J}} \approx 10^6$$

The amount of work done is about one million times smaller than the amount of heat added to the cube.

20-37.   (a) The lengths of the brass rod and bob will both increase with temperature as $\Delta L = \alpha L \Delta T$, so

the fractional change in length is $\dfrac{\Delta L}{L} = \alpha \Delta T = \left(19 \times 10^{-6}/^{\circ}C\right) 20^{\circ}C = \underline{3.8 \times 10^{-4}}$

The period of the pendulum is related to its length as $P = 2\pi\sqrt{\dfrac{L}{g}}$, so the ratio of the periods after

heating to the initial value is

$$\frac{P_2}{P_1} = \sqrt{\frac{L_2}{L_1}} = \sqrt{\frac{L_1 + \Delta L}{L_1}} = \sqrt{1 + \frac{\Delta L}{L_1}} = 1 + \frac{1}{2}\left(\frac{\Delta L}{L_1}\right)$$

The fractional increase in period is $\Delta P = 0.5(3.8 \times 10^{-4}) = \underline{1.9 \times 10^{-4}}$.

(b) Over the course of 1 day (86 400 s), the pendulum clock loses, $\Delta t = (1.9 \times 10^{-4})(86\ 400\ s)$
$= \underline{16\ s}$

20-39.   The center of mass of the mercury is located a distance $l - h/2$ from the pivot point of the pendulum. When the temperature of the system changes, the center of mass changes to a new position $l' - h'/2$. The stated requirement is that the distance of the center of mass with respect to the pivot should remain unchanged, therefore

$l - h/2 = l' - h'/2$ (i)

If the temperature changes by ΔT, the length of the brass rod changes to

$l' = l(1 + \alpha \Delta T)$ (ii)

and the volume of the mercury changes to

$V' = V(1 + \beta \Delta T)$ (iii)

The initial volume of the mercury is $V = hA$ where h is the height of the mercury and A is the diameter of the cylinder. Since the expansion of the glass cylinder is negligible, the height increases with temperature as

$h' = h(1 + \beta \Delta T)$ (iv)

Substituting equations (ii) and (iv) into equation (i) gives

$l - \frac{1}{2}h = l(1 + \alpha \Delta T) - \frac{1}{2}h(1 + \beta \Delta T)$

$l\alpha \Delta T = \frac{1}{2}h\beta \Delta T$

$h = \dfrac{2\alpha}{\beta}l$

20-43.   The rate of heat conducted through the window is

$$\frac{\Delta Q}{\Delta t} = kA\frac{\Delta T}{\Delta x} = (1.0\ J/(s\cdot m\cdot{}^{\circ}C))(1.0\ m)(1.5\ m)\left(\frac{39^{\circ}C}{0.0025\ m}\right) = 23\ 400\ W = \underline{23\ 000\ W}$$

The rate of conduction through the wall in example 6 is 1.8×10^3 W. Therefore, the ratio of the heat conducted through the window to that through the walls is

$\dfrac{23400\ W}{1800\ W} = 13$

The rate through the window is thirteen times greater than through the walls.

20-47. The tank has six glass sides: 2(80 cm × 30 cm), 2(50 cm × 30 cm) and 2(80 cm × 50 cm). Since the temperature inside the tank is higher than that outside the tank, there would be a net flow of heat out of the tank into the surrounding environment without the additional heat added to the tank by the internal heater. To maintain the temperature in the tank, the heater input must match the heat conducted through the walls of the tank.

$$\frac{\Delta Q}{\Delta t} = kA\frac{\Delta T}{\Delta x} = k(A_1 + A_2 + A_3)\frac{\Delta T}{\Delta x}$$

$$= (1.0 \text{ J/(s}\cdot\text{m}\cdot{}^\circ\text{C}))2\left[\begin{array}{l}(0.80 \text{ m} \times 0.30 \text{ m}) + (0.50 \text{ m} \times 0.30 \text{ m}) \\ + (0.80 \text{ m} \times 0.50 \text{ m})\end{array}\right]\frac{(26-18)^\circ\text{C}}{0.003 \text{ m}}$$

$$= 4.2 \times 10^3 \text{ W}$$

20-49. Since heat is conducted through each slab at a constant rate,

$$\frac{\Delta Q}{\Delta t} = k_i A\left(\frac{dT}{dx}\right)_i$$

for each slab designated by i. Therefore, the temperature gradient across the ith slab is

$$\left(\frac{dT}{dx}\right)_i = \frac{\Delta Q}{k_i A\Delta t}$$

Let the temperature difference across the ith slab be ΔT_i. Then, the total temperature difference across n slabs will be

$$\Delta T = \sum_{i=1}^{n}\Delta T_i = \sum_{i=1}^{n}\left(\frac{dT}{dx_i}\right)\Delta x_i = \sum_{i=1}^{n}\frac{\Delta Q}{k_i A\Delta t}\Delta x_i = \frac{\Delta Q}{A\Delta t}\sum_{i=1}^{n}\frac{\Delta x_i}{k_i}$$

which can be rewritten

$$\frac{\Delta Q}{\Delta t} = \frac{A\Delta T}{\sum_{i=1}^{n}\frac{\Delta x_i}{k_i}}$$

20-53. The rate of heat conduction, $\Delta Q/\Delta t$, must be constant through any imaginary cylindrical, coaxial surface of radius r between r_1 and r_2. For the cable length l, the surface area of these imaginary cylinders is $A(r) = 2\pi rl$. For an infinitesimally small radial distance dr, the conduction equation is

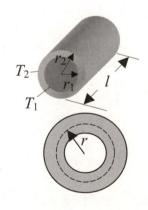

$$\frac{\Delta Q}{\Delta t} = kA\frac{dT}{dr} = k(2\pi rl)\frac{dT}{dr}$$

or

$$\left(\frac{\Delta Q}{\Delta t}\right)\frac{dr}{r} = 2\pi kl \ dT$$

Integrate both sides of the above equation to obtain,

$$\frac{\Delta Q}{\Delta t}\int_{r_1}^{r_2}\frac{dr}{r} = 2\pi kl\int_{T_1}^{T_2} dT$$

$$\frac{\Delta Q}{\Delta t}(\ln r_2 - \ln r_1) = 2\pi kl(T_2 - T_1)$$

$$\frac{\Delta Q}{\Delta t} = \frac{2\pi kl(T_2 - T_1)}{\ln(r_2/r_1)}$$

20-55. Let the rate of growth be $\delta x/\delta t$. For an area A, increasing the thickness by δx means that the volume added per time δt, is $\delta V = A\delta x$. To bring about this phase change requires the removal of heat $\Delta Q = \rho V L_F = \rho(A\delta x) L_F$, where ρ is the density and L_F is the latent heat of fusion for water. The magnitude of the rate of heat transfer is

$$\left|\frac{\Delta Q}{\Delta t}\right| = \rho A L_F \frac{\delta x}{\delta t}$$

This heat transfer is equal to the heat conducted through the ice to the air above the ice from the water below.

$$\left|\frac{\Delta Q}{\Delta t}\right| = kA\frac{\Delta T}{\Delta x} = \rho A L_F \frac{\delta x}{\delta t}$$

$$\frac{\delta x}{\delta t} = \frac{k\Delta T}{\rho L_F \Delta x} = \frac{(1.3 \text{ J/(s}\cdot\text{m}\cdot°\text{C)})(20°\text{C})}{(917 \text{ kg/m}^3)(3.34 \times 10^5 \text{ J/kg})(0.060 \text{ m})}$$

$$= 1.3 \times 10^{-6} \ \tfrac{\text{m}}{\text{s}}\left(\tfrac{100 \text{ cm}}{1 \text{ m}}\right)\left(\tfrac{3600 \text{ s}}{1 \text{ h}}\right) = \underline{0.51 \text{ cm/h}}$$

20-63. The heat that is removed from the cream to freeze it and further cool it goes into vaporizing the liquid nitrogen.

$$\underbrace{m_N L_V}_{\text{heat gained by the nitrogen}} = \underbrace{mc_c\Delta T_C}_{\text{heat lost by the cream}} + \underbrace{mL_F}_{\text{heat lost to freeze}} + \underbrace{mc_I\Delta T_I}_{\text{heat lost by the ice cream}}$$

$$m_N = \frac{m(c_c\Delta T_C + L_F + c_I\Delta T)_I}{L_V}$$

$$= \frac{2.0 \text{ kg}\left[(2900 \text{ J/(kg}\cdot°\text{C)})(10°\text{C}) + 3.34 \times 10^5 \text{ J/kg} + (2200 \text{ J/(kg}\cdot°\text{C)})(10°\text{C})\right]}{2.00 \times 10^5 \text{ J/kg}}$$

$$= \underline{3.9 \text{ kg}}$$

20-69. Since density is defined as $\rho = m/V$. The volume of water evaporated per hour is

$$\frac{V}{t} = \frac{(m/t)}{\rho}$$

The mass evaporated in time t is determined by the amount of heat absorbed by the water in that time:

$$\frac{Q}{t} = \frac{mL_V}{t}$$

Therefore, the volume of the Mediterranean Sea evaporated per hour is

$$\frac{V}{t} = \frac{(m/t)}{\rho} = \frac{(Q/t)}{\rho L_V} = \frac{A(P/A)}{\rho L_V}$$

$$= \frac{2.9 \times 10^6 \text{ km}^2 \left(\tfrac{1000 \text{ m}}{1 \text{ km}}\right)^2 \left(1.0 \times 10^3 \ \tfrac{\text{W}}{\text{m}^2}\right)}{(1000 \text{ kg/m}^3)(2.43 \times 10^6 \text{ J/kg})}$$

$$= 1.2 \times 10^6 \ \tfrac{\text{m}^3}{\text{s}}\left(\tfrac{3600 \text{ s}}{1 \text{ h}}\right)\left(\tfrac{1 \text{ km}}{1000 \text{ m}}\right)^3 = \underline{4.3 \text{ km}^3/\text{h}}$$

20-71. The air conditioner takes in and sends out air at constant pressure, $p = 1.01 \times 10^5$ N/m². Let's assume that since air is mostly nitrogen that its specific heat is approximately equal to that of nitrogen, $c_p = 29.1$ J/(mol • K). The heat absorbed by the air is $Q = nc_p\Delta T$, where n is the number of moles, which can be determined using the Ideal-Gas Law,

$$pV = nRT$$

$$n = \frac{pV}{RT}$$

where $T = 30°C$ is the initial temperature of the air. Therefore, combining this result with the equation for Q/t,

$$Q = \frac{pV}{RT}c_p\Delta T$$

$$\Delta T = \frac{QRT}{pVc_p} = \frac{(Q/t)RT}{p(V/t)c_p} = \frac{\left(1.2 \times 10^7 \text{ J/h}\right)\left(\frac{1\text{ h}}{60\text{ min}}\right)\left(8.31 \text{ J/(K•mol)}\right)\left(303 \text{ K}\right)}{\left(1.01 \times 10^5 \text{ N/m}^2\right)\left(15 \frac{\text{m}^3}{\text{min}}\right)\left(29.1 \text{ J/(K•mol)}\right)} = 11.4 \text{ K}$$

The temperature of the air exiting the fan is $T_f = T + \Delta T = 30°C + 11.4°C = \underline{41°C}$.

20-73. One kg of helium is equal to

$$n = \frac{1000 \text{ g}}{4.00260 \text{ g/mole}} = 249.838 \text{ moles}$$

The specific heat of one kilogram of helium is

$$C_{V(kg)} = nC_{V(mole)} = \frac{249.838 \text{ moles}}{1 \text{ kg}}(12.5 \text{ J/(mole•K)}) = \underline{3.12 \times 10^3 \text{ J/(kg•K)}}$$

The above calculation may then be repeated for the other gases in Table 20.5 to create the following table of results.

Gas	Molecular weight (g)	Moles/kg	C_V (J/(kg•K))
He	4.00260	249.838	3.12×10^3
Ar	39.948	25.033	3.13×10^2
N$_2$	28.0134	35.697	7.42×10^2
O$_2$	31.999	31.251	6.50×10^2
CO	28.0104	35.701	7.39×10^2
NH$_3$	17.0305	58.718	1.60×10^3
CH$_4$	16.04276	62.333	1.69×10^3

The gas with the highest specific heat per kilogram is helium, while the lowest is for argon.

20-81. Each hour that you breathe in the cold dry air, you add to it

$$\frac{\Delta m_{vapor}}{\Delta t} = \left(\frac{0.041 \text{ kg vapor}}{1 \text{ kg air}}\right)\left(\frac{0.45 \text{ kg air}}{1 \text{ h}}\right) = 0.1845 \text{ kg vapor/h}$$

The heat required to warm and vaporize the air is

$$\frac{\Delta Q}{\Delta t} = \frac{\Delta m}{\Delta t}C_p\Delta T + \frac{\Delta m}{\Delta t}L_v$$

$$= \left(\frac{0.45 \text{ kg}}{1 \text{ h}} \right) \left(\frac{1 \text{ h}}{3600 \text{ s}} \right) \left(1000 \text{ J/(kg} \cdot {}^\circ\text{C)} \right) \left(37{}^\circ\text{C} - (-30{}^\circ\text{C}) \right)$$
$$+ \left(0.1845 \text{ kg vapor/h} \right) \left(2.42 \times 10^6 \text{ J/kg} \right)$$
$$= \left(4.46 \times 10^5 \text{ J/h} \right) \left(\frac{1 \text{ kcal}}{4187 \text{ J}} \right) = \underline{110 \text{ kcal/h}}$$

20-85. (a) From the ideal gas law,

$$V = \frac{nRT}{p} = \frac{(1.0 \text{ kg}) \left(\frac{1 \text{ mole}}{0.028 \text{ kg}} \right) \left(8.31 \text{ J/(mole} \cdot \text{K)} \right) (293 \text{ K})}{1.2 \times 10^6 \text{ N/m}^2} = 0.072 \text{ m}^3$$

(b) Since $pV^\gamma = $ constant, we have

$$p_1 V_1^\gamma = p_2 V_2^\gamma$$

$$V_2^\gamma = V_1^\gamma \left(\frac{p_1}{p_2} \right)$$

$$V_2 = V_1 \left(\frac{p_1}{p_2} \right)^{1/\gamma} = 0.072 \text{ m}^3 \left(\frac{1.2 \times 10^6 \text{ N/m}^2}{1.01 \times 10^5 \text{ N/m}^2} \right)^{1/1.40} = \underline{0.42 \text{ m}^3}$$

The Ideal-Gas Law is $pV = nRT$ or $p = nRT/V$; and since $pV^\gamma = $ constant, we have

$$nRT \frac{V^\gamma}{V} = \text{ constant}$$

$$TV^{\gamma-1} = \text{ constant}$$

Applying this to this situation gives,

$$T_1 V_1^{\gamma-1} = T_2 V_2^{\gamma-1}$$

$$T_2 = T_1 \left(\frac{V_1}{V_2} \right)^{\gamma-1} = 293 \text{ K} \left(\frac{0.072 \text{ m}^3}{0.42 \text{ m}^3} \right)^{1.40-1} = \underline{145 \text{ K or } -128{}^\circ\text{C}}$$

20-93. Assuming the ship is composed of steel, the change in length of the ship will be

$$\Delta L = \alpha L \Delta T = (12 \times 10^{-6}/{}^\circ\text{C})(458 \text{ m})[40{}^\circ\text{C} - (-20{}^\circ\text{C})] = \underline{0.33 \text{ m}}$$

Similarly, the width of the ship will increase by

$$\Delta w = \alpha w \Delta T = (12 \times 10^{-6}/{}^\circ\text{C})(69 \text{ m})[40{}^\circ\text{C} - (-20{}^\circ\text{C})] = \underline{0.050 \text{ m}}$$

As a result of this thermal expansion, the increase in area is

$$\Delta A = (L + \Delta L)(w + \Delta w) - Lw = (458.33 \text{ m})(69.050 \text{ m}) - (458 \text{ m})(69 \text{ m}) = \underline{46 \text{ m}^2}$$

20-97. The ice-water mixture is at $0{}^\circ\text{C}$ before the dry ice (frozen CO_2 at $-79{}^\circ\text{C}$). For water of mass m to become ice an amount of heat $Q = mL_\text{F}$ must be removed from the ice-water mixture. This heat comes from the evaporation of the 0.30 kg of dry ice as it is added to the ice-water. The heat required to vaporize the dry ice is

$$Q = m_{\text{dry ice}} L_\text{F} = \left(0.30 \text{ kg} \right) \left(5.80 \times 10^5 \text{ J/kg} \right) = 1.74 \times 10^5 \text{ J}$$

This amount of heat removed from the water causes the following mass of water to freeze to ice at $0{}^\circ\text{C}$.

$$m = \frac{Q}{L_\text{F}} = \frac{1.74 \times 10^5 \text{ J}}{3.34 \times 10^5 \text{ J/kg}} = \underline{0.52 \text{ kg}}$$

21-1. The change in internal energy for this ideal monatomic gas is the difference between the heat absorbed Q by the gas and the work that it does on the piston. The heat input into the system is
$$Q = nC_p\Delta T = (1.0 \text{ mole})(20.8 \text{ J/(mole} \cdot \text{K}))(90 \text{ K}) = \underline{1.9 \times 10^3 \text{ J}}$$
where c_p is given in Table 20.5. The change in the internal energy is then,
$$\Delta E = Q - W = nC_p\Delta T - W$$
$$= (1.0 \text{ mole})(20.8 \text{ J/(mole} \cdot \text{K}))(90 \text{ K}) - 800 \text{ J} = \underline{1.1 \times 10^3 \text{ J}}$$

21-3. (a) In step **A**, the gas isovolumetrically ($\Delta V = 0$) expands, so the work done by the gas $W = p\Delta V = \underline{0 \text{ J}}$
The heat absorbed by the ideal gas is $Q = nc_V\Delta T$. From the ideal gas law, we have
$$p_1V_1 = nRT_1 \quad \text{or} \quad V_1 = \frac{nRT_1}{p_1} = \frac{nRT_2}{p_2}$$
since the volume is constant. Solving for $\Delta T = T_2 - T_1$, gives
$$\Delta T = T_1\left(\frac{p_2}{p_1} - 1\right)$$

The system is initially at standard temperature and pressure (STP), $T_1 = 273$ K and $P_1 = 1.0$ atm. One mole of an ideal gas at STP has a volume of 22.4 liters. The heat absorbed in step **A** is
$$Q = nC_V T_1\left(\frac{p_2}{p_1} - 1\right) = 4.0 \text{ liters}\left(\tfrac{1 \text{ mole}}{22.4 \text{ liters}}\right)(12.5 \text{ J/(mole} \cdot \text{K}))(273 \text{ K})\left(\frac{2p_1}{p_1} - 1\right) = \underline{610 \text{ J}}$$

The internal energy in this case is equal to the heat absorbed by the system,
$$\Delta E = Q - W = Q = \underline{610 \text{ J}}$$
(b) In step **B**, the gas isobarically ($\Delta P = 0$) expands by $\Delta V = 4.0$ liters $(10^{-3} \text{ m}^3/\text{liter}) = 4.0 \times 10^{-3}$ m^3, so the work done by the gas is
$$W = p\Delta V = (2.02 \times 10^5 \text{ N/m}^2)(4.0 \times 10^{-3} \text{ m}^3) = 808 \text{ J} = \underline{810 \text{ J}}$$
The amount of heat absorbed by the system in this process is again $Q = nC_V\Delta T$. From the ideal gas law, we have
$$p_2V_1 = nRT_2 \quad \text{and} \quad p_2V_2 = nRT_3$$
$$T_2 = \frac{p_2V_1}{nR} \quad \text{and} \quad T_3 = \frac{p_2V_2}{nR}$$
$$\Delta T = T_3 - T_2 = \frac{p_2V_2}{nR} - \frac{p_2V_1}{nR} = \frac{p_2}{nR}\Delta V$$
$$= \frac{2.02 \times 10^5 \text{ N/m}^2}{4.0 \text{ liters}\left(\tfrac{1 \text{ mole}}{22.4 \text{ liters}}\right)(8.31 \text{ J/(mole} \cdot \text{K}))}\left(4.0 \times 10^{-3} \text{ m}^3\right) = 544.5 \text{ K}$$

Then,
$$Q = nC_p\Delta T = 4.0 \text{ liters}\left(\tfrac{1 \text{ mole}}{22.4 \text{ liters}}\right)(20.8 \text{ J/(mole} \cdot \text{K}))(544.5 \text{ K}) = 2022 \text{ J} = \underline{2.0 \times 10^3 \text{ J}}$$

The internal energy in this case is
$$\Delta E = Q - W = Q = 2.0 \times 10^3 \text{ J} - 810 \text{ J} = \underline{1.2 \times 10^3 \text{ J}}$$

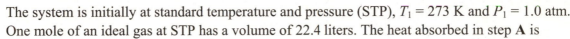

21-5.   In an adiabatic process, heat is neither absorbed nor emitted from a system; $Q = 0$. Since work is being done on the gas instead of by the gas, the internal energy of the system increases
$$\Delta E = Q - W = -W$$

Since work is done on the gas, the work done has a negative value and the internal energy is increasing in this situation. According to section 20.6, the change in internal energy is also equal to $nC_V\Delta T$. Therefore,

$$\Delta E = nC_V\Delta T = (0.0300 \text{ mole})(20.8 \text{ J/(mole} \cdot \text{K)})((790 - 40) \text{ K}) = 470 \text{ J}$$

The work done is $W = \underline{-470 \text{ J}}$.

21-7.   (a) Assuming the gas obeys the ideal gas law, $pV = nRT$,

$$n = \frac{pV}{RT} = \frac{(1.01 \times 10^5 \text{ N/m}^2)(0.100 \text{ m}^3)}{(8.31 \text{ J/(mole} \cdot \text{K)})(283 \text{ K})} = \underline{4.29 \text{ moles}}$$

(b) The work done by the gas as it expands is

$$W = p\Delta V = (1.01 \times 10^5 \text{ N/m}^2)((0.110 - 0.100)\text{m}^3) = \underline{1010 \text{ J}}$$

(c) The change in internal energy is $\Delta E = Q - W = 3500 \text{ J} - 1010 \text{ J} = \underline{2490 \text{ J}}$

(d) Consider section 20.5 and equations 20.30 to 20.33, in particular. The internal energy of the gas is $\Delta E = fnR\Delta T$, where f is a fraction related to the nature of the gas (monatomic or diatomic). For the situation given,

$$f = \frac{\Delta E}{nR\Delta T} = \frac{2490 \text{ J}}{(4.29 \text{ moles})(8.31 \text{ J/(mole} \cdot \text{K)})(28 \text{ K})} = \frac{5}{2}$$

Therefore, $C_V = 5R/2$ and the gas is $\underline{\text{diatomic}}$.

21-11.   The work done during this process is
$$W = p\Delta V$$
$$= 0.25 \text{ atm } (1.01 \times 10^5 \text{ N/m}^2/1.0 \text{ atm})(0.5 \times 10^{-3} \text{ m}^3 - 2.0 \times 10^{-3} \text{ m}^3) = \underline{-37.9 \text{ J}}$$
During the process 75 J flows out of the system, therefore $Q = -75$ J. The change in internal energy is $\Delta E = Q - W = -75 \text{ J} - (-37.9 \text{ J}) = \underline{-37.1 \text{ J}}$

21-13.   In the temperature range of 5°C to 30°C, ethanol is a liquid. The amount of heat added to the ethanol is
$$Q = mc\Delta T$$

According to Table 20.1, the specific heat of ethanol at 20°C is 2430 J/(kg • °C). To determine the mass of the liquid, we need to know its density, which is given for 20°C. Therefore, we need to know the volume of the ethanol at 20°C. At 5.0°C, the volume is 1.00 liter.

As the alcohol is heated, it undergoes volumetric thermal expansion

$$\Delta V = \beta V\Delta T = (1.01 \times 10^{-3}/°\text{C})(1.00 \text{ liter})\left(\frac{1.00 \times 10^{-3} \text{ m}^3}{1.00 \text{ liter}}\right)(20°\text{C} - 5°\text{C}) = 1.51 \times 10^{-5} \text{ m}^3$$

where β is the volumetric thermal expansion coefficient listed in Table 20.2. The volume at 20°C is

$$V + \Delta V = (1.00 \times 10^{-3} \text{ m}^3) + (1.51 \times 10^{-5} \text{ m}^3) = (1.015 \times 10^{-3} \text{ m}^3)$$

The mass of the ethanol is therefore,
$$m = \rho V = (789 \text{ kg/m}^3)(1.015 \times 10^{-3} \text{ m}^3) = 0.801 \text{ kg}$$

Then,
$$Q = mc\Delta T = (0.801 \text{ kg})(2430 \text{ J/(kg} \cdot °\text{C)})(30°\text{C} - 5°\text{C}) = \underline{4.87 \times 10^4 \text{ J}}$$

The volume expansion between 5°C and 30°C is

$\Delta V = \beta V \Delta T = (1.01 \times 10^{-3} /°C)(1.00 \text{ liter}) \left(\frac{1.00 \times 10^{-3} \text{ m}^3}{1.00 \text{ liter}} \right)(30°C - 5°C) = 2.53 \times 10^{-5} \text{ m}^3$

The work done by the alcohol in its expansion on the surrounding air is

$W = p\Delta V = 1.00 \text{ atm } (1.01 \times 10^5 \text{ N/m}^2/1.00 \text{ atm})(2.53 \times 10^{-5} \text{ m}^3) = \underline{2.56 \text{ J}}$

21-15. This is an example of an isothermal process, $\Delta T = 0$ K. Since the temperature does not change, there is no change in the internal energy of the ideal gas; and $W = Q$. The work done by the gas in its expansion, as determined in problem 21-10(a), is

$W = nRT \ln \frac{V_2}{V_1} = (10 \text{ moles})(8.31 \text{ J/(mole} \cdot \text{K)})(323 \text{ K}) \ln \left(\frac{10 \text{ liters}}{2 \text{ liters}} \right) = \underline{4.3 \times 10^4 \text{ J}}$

21-17. The temperature may be determined from the Ideal-Gas Law,

$pV = nRT$

$T = \frac{p_2 V_2}{nR} = \frac{(1.01 \times 10^5 \text{ N/m}^2)(0.10 \text{ m}^3)}{(8.31 \text{ J/(mole} \cdot \text{K)})(5.00 \text{ moles})} = 243 \text{ K}$

Because this is an isothermal process, $\Delta T = 0$ K; and there is no change in the internal energy of the ideal gas; and the work done as the gas expands is

$W = Q = nRT \ln \frac{V_2}{V_1}$

from which the initial volume may be determined since the quantity of work done by the gas is given.

$\frac{W}{nRT} = \ln \frac{V_2}{V_1}$

$e^{\frac{W}{nRT}} = \frac{V_2}{V_1}$

$V_1 = V_2 e^{-\frac{W}{nRT}} = (0.10 \text{ m}^3) e^{-(2.0 \times 10^4 \text{ J})/[(5.00 \text{ moles})(8.31 \text{ J/(mole} \cdot \text{K)})(243 \text{ K})]} = \underline{0.014 \text{ m}^3}$

The initial pressure was

$p_1 = \frac{nRT}{V_1} = \frac{(5.00 \text{ moles})(8.31 \text{ J/(mole} \cdot \text{K)})(243 \text{ K})}{0.014 \text{ m}^3} = \underline{7.2 \times 10^5 \text{ N/m}^2}$

21-19. The diagram shows the indirect process described in the problem. The net work done against the atmosphere is the same in the indirect process as in the direct process since the changes in the volume are the same. The net changes in the internal energies is also the same. Because the work done and the internal energy are the same, then the heat must also be the same. For 1.0 kg of water, the latent heat per kilogram is

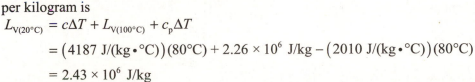

$L_{V(20°C)} = c\Delta T + L_{V(100°C)} + c_p \Delta T$

$= (4187 \text{ J/(kg} \cdot °\text{C)})(80°C) + 2.26 \times 10^6 \text{ J/kg} - (2010 \text{ J/(kg} \cdot °\text{C)})(80°C)$

$= \underline{2.43 \times 10^6 \text{ J/kg}}$

21-23.   The work done by the runner is $W = \Delta U = mgh$. So, the rate that work is being done is
$$\frac{\Delta W}{\Delta t} = mg \frac{\Delta h}{\Delta t} = mgv$$

The efficiency is determined by the ratio of this work to the total energy generated by the body that includes the work done plus the waste heat generated.

$$e = \frac{(\Delta W/\Delta t)}{(\Delta W/\Delta t) + (\Delta Q/\Delta t)} = \frac{mgv}{mgv + (\Delta Q/\Delta t)}$$

$$= \frac{(70 \text{ kg})(9.81 \text{ m/s}^2)(0.30 \text{ m/s})}{(70 \text{ kg})(9.81 \text{ m/s}^2)(0.30 \text{ m/s}) + (1300 \text{ W})} = \underline{0.14 \text{ or } 14\%}$$

21-25.   According to equation 21.5, $e = \dfrac{W}{Q_1} = 1 - \dfrac{Q_2}{Q_1}$, the efficiency may be determined in more than

one way. In this case, we are given Q_1 and Q_2 and asked to find the efficiency and the work done by the engine.

$$e = 1 - \frac{Q_2}{Q_1} = 1 - \frac{8.0 \times 10^6 \text{ J}}{2.0 \times 10^7 \text{ J}} = \underline{0.60 \text{ or } 60\%}$$

The work done is $W = Q_1 - Q_2 = \underline{1.2 \times 10^7 \text{ J}}$

21-31.   The engines each do 1100 hp of work per second and the efficiency is 20%, so the heat input to the system to deliver this work is somewhat greater. To find that rate of heat input, remember that efficiency is defined as $e = W/Q_1$. Dividing both factors on the right side of the equation by t gives the ratio of the rate of work to the rate of heat input.

$$e = \frac{W/t}{Q_1/t}$$

$$\frac{Q_1}{t} = \frac{W/t}{e} = \frac{2200 \text{ hp}\left(\frac{745.7 \text{ W}}{1 \text{ hp}}\right)}{0.20} = 8.2 \times 10^6 \text{ W} = \underline{8.2 \times 10^6 \text{ J/s}}$$

The rate of fuel input needed to supply this heat input is

$$\frac{m}{t} = \frac{Q/t}{E/m} = \frac{8.2 \times 10^6 \text{ J/s}}{4.4 \times 10^7 \text{ J/kg}} = \underline{0.19 \text{ kg/s}}$$

21-33.   The efficiency is determined by the temperatures of the high temperature (T_1) and low temperature (T_2) reservoirs.

$$e = 1 - \frac{T_2}{T_1} = 1 - \frac{300 \text{ K}}{1 \times 10^{11} \text{ K}} = 1 - (3 \times 10^9) = \underline{0.999\ 999\ 997}$$

This is very close to 100% efficiency. However, working with such high temperatures can be a difficult engineering problem in practice.

21-35.   The efficiency is determined by the temperatures of the high temperature (T_1) and low temperature (T_2) reservoirs. The efficiency e is the ratio of the work done W by the engine to the amount of heat absorbed Q_1. Therefore,

$$e = \frac{W}{Q_1} = 1 - \frac{T_2}{T_1}$$

$$Q_1 = \frac{W}{1 - \dfrac{T_2}{T_1}} = \frac{5.0 \times 10^4}{1 - \left(\dfrac{273 \text{ K}}{373 \text{ K}}\right)} = \underline{1.9 \times 10^5 \text{ J}}$$

The amount of waste heat produced is the difference between the heat absorbed and the work done,

$$Q_2 = Q_1 - W = (1.9 \times 10^5 \text{ J}) - (5.0 \times 10^4 \text{ J}) = \underline{1.4 \times 10^5 \text{ J}}$$

21-37. According to equation 21.5, $e = \dfrac{W}{Q_1} = 1 - \dfrac{Q_2}{Q_1}$, the efficiency may be determined in more than one way. In this case, we are given Q_1 and Q_2 and asked to find the efficiency and the mechanical power output (the work done per second) the engine. Therefore,

$$e = 1 - \frac{Q_2}{Q_1} = 1 - \frac{Q_2/t}{Q_1/t} = 1 - \frac{1400 \text{ J/s}}{2500 \text{ J/s}} = \underline{0.44 \text{ or } 44\%}$$

The power output is

$$e = \frac{W}{Q} = \frac{W/t}{Q_1/t}$$

$$P = \frac{W}{t} = e(Q_1/t) = (0.44)(2500 \text{ J/s}) = 1100 \text{ J/s} = \underline{1100 \text{ W}}$$

21-39. The flowchart for the operation of a heat pump is shown in Figure 21.16. Heat is taken from the inside of the house at T_2 and sent to the high temperature reservoir at T_1 outside through the work done by heat pump unit. The efficiency of the heat pump is

$$e = \frac{W}{Q_1} = 1 - \frac{Q_2}{Q_1} = 1 - \frac{T_2}{T_1}$$

To apply the thermodynamic relationships, temperatures must be converted into degrees Kelvin. Each °F = (5/9)°C. Since the freezing point of water is at 32°F, the temperatures given are:

$$T_2 = \left(75°F - 32°F\right)\left(\frac{5°C}{9°F}\right) + 273°C = 297 \text{ K}$$

and

$$T_1 = \left(104°F - 32°F\right)\left(\frac{5°C}{9°F}\right) + 273°C = 313 \text{ K}$$

Therefore, the heat transferred to the hot reservoir by the heat pump operating in cooling mode each hour will be

$$\frac{Q_2}{Q_1} = \frac{T_2}{T_1}$$

$$Q_1 = Q_2\left(\frac{T_1}{T_2}\right) = (5.0 \times 10^6 \text{ J})\left(\frac{313 \text{ K}}{297 \text{ K}}\right) = 5.27 \times 10^6 \text{ J}$$

Therefore, the work input to the heat pump is

$$W = Q_1 - Q_2 = 5.27 \times 10^6 \text{ J} - 5.0 \times 10^6 \text{ J} = 2.7 \times 10^5 \text{ J}$$

The rate of work done is

$$P = \frac{W}{t} = \frac{2.7 \times 10^5 \text{ J}}{1 \text{ h}}\left(\frac{1 \text{ h}}{3600 \text{ s}}\right) = \underline{75 \text{ W}}$$

21-41. When a heat pump is used to provide heat to a home, the cofficient of performance (CP) is defined as

$$\text{CP} = \frac{\text{energy transferred to the high temperature reservoir}}{\text{work done by the heat pump}} = \frac{Q_1}{W}$$

The efficiency of a carnot engine is

$$e = \frac{W}{Q_1} = 1 - \frac{Q_2}{Q_1} = 1 - \frac{T_2}{T_1} = \frac{T_1 - T_2}{T_1}$$

Therefore,

$$\text{CP} = \frac{Q_1}{W} = \frac{T_1}{T_1 - T_2} = \frac{293 \text{ K}}{293 - 278 \text{ K}} = \underline{19.5}$$

This value represents the maximum CP for a system operating between these two temperatures. In practice, the CP values for real heat pumps is somewhat smaller than this value.

21-45. (a) The work done during each step for the ideal monatomic gas is $W = p\Delta V$. Therefore, the work done during step 1 is zero J, since there is no change in volume. During step 2, the work done is

$$W_2 = p_2(V_2 - V_1) = (1 \times 10^5 \text{ N/m}^2)(0.03 \text{ m}^3 - 0.01 \text{ m}^3) = 2 \times 10^3 \text{ J}$$

In step 3, both the pressure and the volume change, such processes may be either isothermal or adiabatic. To determine which type of process occurs in this system, consider the beginning and ending points of step 1, applying the Ideal-Gas Law

$$P_1 V_1 = nRT_1$$

$$T_1 = \frac{P_1 V_1}{nR} = \frac{(3 \times 10^5 \text{ N/m}^2)(0.01 \text{ m}^3)}{(1.00 \text{ mole})(8.31 \text{ J/(mole} \cdot \text{K)})} = 361 \text{ K}$$

$$\frac{P_1}{T_1} = \frac{P_2}{T_2}$$

$$T_2 = \frac{P_2}{P_1} T_1 = \tfrac{1}{3} T_1 = \tfrac{1}{3}(361 \text{ K}) = 120 \text{ K}$$

Then, in step 2, there is an isobaric expansion, the temperature at the beginning of step 3 is

$$\frac{V_2}{T_2} = \frac{V_3}{T_3}$$

$$T_3 = \frac{V_3}{V_2} T_2 = 3T_2 = 3\left(\tfrac{1}{3}T_1\right) = T_1$$

The temperature in step 3 does not change, so the process is isothermal. The work done in an isothermal process is

$$W_3 = p_2 V_2 \ln \frac{V_3}{V_2} = (1 \times 10^5 \text{ N/m}^2)(0.03 \text{ m}^3) \ln \left(\frac{0.01 \text{ m}^3}{0.03 \text{ m}^3} \right) = -3.3 \times 10^3 \text{ J}$$

The total work done during one cycle is

$$W = W_1 + W_2 + W_3 = 0 + 2.0 \times 10^3 \text{ J} - 3.3 \times 10^3 \text{ J} = \underline{-1.3 \times 10^3 \text{ J}}$$

The negative value of the net work indicates that work is done on the gas during the cycle.

(b) During the isobaric step 2, the heat flow $Q = mc_p\Delta T$. Since the change in temperature is positive, there is a net flow of heat into the system. Heat is absorbed during step 2. During isothermal step 3, the heat flow is equal to the amount of work, since the work is negative, the heat flow is again negative and there is a net heat flow out of the system. Heat is rejected in step 3.

(c) During isovolumetric step 1, heat is rejected by the system, since $Q = mc_p\Delta T$ and T_1 is greater than T_2.

(d) Just as in single step processes that occur between high and low temperature reservoirs, the efficiency is still defined in a cyclical processes as the ratio of the work done by the system to the quantity of heat rejected to the high temperature reservoir. Heat is rejected to the high temperature reservoir in step 3.

$$e = \frac{W}{Q_3} = \frac{W}{W_3} = \frac{W}{W_3} = \frac{-1.3 \times 10^3 \text{ J}}{-3.3 \times 10^3 \text{ J}} = \underline{0.39}$$

21-49.   (a) For the carnot heat pump operating in cooling mode, the heat rejected each hour to the high temperature reservoir is found from

$$\frac{Q_2}{Q_1} = \frac{T_2}{T_1}$$

$$Q_1 = Q_2\left(\frac{T_1}{T_2}\right) = (8.4 \times 10^6 \text{ J})\left(\frac{300 \text{ K}}{294 \text{ K}}\right) = 8.57 \times 10^6 \text{ J}$$

Therefore, the work input to the heat pump each hour is

$W = Q_1 - Q_2 = 8.57 \times 10^6 \text{ J} - 8.4 \times 10^6 \text{ J} = 1.71 \times 10^5 \text{ J}$

The rate of work done is

$$P = \frac{W}{t} = \frac{1.71 \times 10^5 \text{ J}}{1 \text{ h}}\left(\frac{1 \text{ h}}{3600 \text{ s}}\right) = \underline{48 \text{ W}}$$

(b) The air conditioner unit requires 950 W of work input, which is somewhat more than the heat pump from part (a).

$$\text{factor} = \frac{P_{ac}}{P_{hp}} = \frac{950 \text{ W}}{48 \text{ W}} = \underline{20}$$

21-51.   (a) $e_{\text{turbine}} = 1 - \frac{T_2}{T_1} = 1 - \frac{260 + 273 \text{ K}}{540 + 273 \text{ K}} = \underline{0.34}$

$e_{\text{engine}} = 1 - \frac{T_3}{T_2} = 1 - \frac{38 + 273 \text{ K}}{260 + 273 \text{ K}} = \underline{0.42}$

(b) The maximum net efficiency for the two components would occur if all of the heat that is rejected from the turbine goes into the stream engine. In practice, heat would be lost to the surroundings. Here, we'll calculate the maximum efficiency,

$$e = \frac{W_{\text{net}}}{Q_1} = \frac{W_{\text{turbine}} + W_{\text{engine}}}{Q_1} = \frac{e_{\text{turbine}}Q_1 + e_{\text{engine}}Q_2}{Q_1}$$

$$= e_{\text{turbine}} + e_{\text{engine}}\left(\frac{Q_2}{Q_1}\right) = e_{\text{turbine}} + e_{\text{engine}}\left(\frac{T_2}{T_1}\right) = 0.34 + 0.42\left(\frac{260 + 273 \text{ K}}{540 + 273 \text{ K}}\right) = \underline{0.62}$$

If a single engine were utilized between the high and low temperature reservoirs,

$$e = 1 - \frac{T_3}{T_1} = 1 - \frac{38 + 273 \text{ K}}{540 + 273 \text{ K}} = \underline{0.62}$$

The two efficiencies are the same.

21-53.   The change in entropy as the water vaporizes is

$$\Delta S_1 = \int \frac{dQ}{T} = \frac{\Delta Q}{T} = \frac{mL_V}{T} = \frac{(1.0 \text{ kg})(2.26 \times 10^6 \text{ J/kg})}{373 \text{ K}} = 6.06 \times 10^3 \text{ J/K}$$

In addition, the water vapor expands as the vaporization occurs and does work against the surrounding air. For the expanding gas, $dQ = dW = pdV$, so the change in entropy due to the expansion is

$$\Delta S_2 = \int \frac{dQ}{T} = \int \frac{pdV}{T} = \int nR \frac{dV}{V} = nR \ln \frac{V_2}{V_1}$$

The initial volume of the water, assuming thermal expansion is negligible, is
$V_1 = m/\rho = 1.0 \text{ kg}/(1000 \text{ kg/m}^3) = 0.0010 \text{ m}^3$
The volume of 1.0 kg of water vapor at 1.0 atm is

$$V_2 = \frac{nRT}{p} = \frac{(1.0 \text{ kg})\left(\frac{1 \text{ mole}}{0.018 \text{ kg}}\right)(8.31 \text{ J/(mole} \cdot \text{K}))(373 \text{ K})}{1.01 \times 10^5 \text{ N/m}^2} = 1.705 \text{ m}^3$$

Therefore,

$$\Delta S_2 = nR \ln \frac{V_2}{V_1} = (1.0 \text{ kg})\left(\frac{1 \text{ mole}}{0.018 \text{ kg}}\right)(8.31 \text{ J/(mole} \cdot \text{K})) \ln \frac{1.705 \text{ m}^3}{0.001 \text{ m}^3} = 3.44 \times 10^3 \text{ J/K}$$

The net change in entropy is $\Delta S = \Delta S_1 + \Delta S_2 = 6.06 \times 10^3 \text{ J/K} + 3.44 \times 10^3 \text{ J/K} = \underline{9.5 \times 10^3 \text{ J/K}}$

21-55.   From problem 49, the rate of heat removed from the inside is

$$\frac{\Delta Q_2}{\Delta t} = 8.4 \times 10^6 \frac{\text{J}}{\text{h}} \left(\frac{1 \text{ h}}{3600 \text{ s}}\right) = 2.33 \times 10^3 \text{ W}$$

The total entropy change is the sum of the entropy change on the inside and outside. The change in heat on the inside is ΔQ_2 where the temperature is T_2; and ΔQ_1 on the outside where the temperature is T_1.

$$\Delta S = \frac{\Delta Q_1}{T_1} - \frac{\Delta Q_2}{T_2}$$

$$\frac{\Delta S}{\Delta t} = \frac{\Delta Q_1/\Delta t}{T_1} - \frac{\Delta Q_2/\Delta t}{T_2} = \frac{2.33 \times 10^3 \text{ W} + 950 \text{ W}}{27 + 273 \text{ K}} - \frac{2.33 \times 10^3 \text{ W}}{21 + 273 \text{ K}} = \underline{3.0 \text{ W/K}}$$

21-61.   The melting temperatures and heats of fusion are given in Table 20.4.

$$\Delta S_{Al} = \frac{\Delta Q}{T_{melt}} = \frac{mL_F}{T_{melt}} = \frac{(1.00 \text{ kg})(3.99 \times 10^5 \text{ J/kg})}{660 + 273 \text{ K}} = \underline{430 \text{ J/K}}$$

$$\Delta S_{Fe} = \frac{mL_F}{T_{melt}} = \frac{(1.00 \text{ kg})(2.7 \times 10^5 \text{ J/kg})}{1535 + 273 \text{ K}} = \underline{150 \text{ J/K}}$$

$$\Delta S_{Ag} = \frac{mL_F}{T_{melt}} = \frac{(1.00 \text{ kg})(9.9 \times 10^4 \text{ J/kg})}{962 + 273 \text{ K}} = \underline{80 \text{ J/K}}$$

$$\Delta S_{Hg} = \frac{mL_F}{T_{melt}} = \frac{(1.00 \text{ kg})(1.1 \times 10^4 \text{ J/kg})}{-39 + 273 \text{ K}} = \underline{47 \text{ J/K}}$$

The change in entropy seems to decrease with increasing atomic number. The largest value occurs for aluminum ($Z = 13$) and the smallest value occurs for mercury ($Z = 80$).

21-63. As the automobile moves, it does work against the various sources of friction and absorbs heat.

$$W = fs = \Delta Q$$

The rate of work done is thus,

$$\frac{W}{t} = f\frac{\Delta x}{\Delta t} = \frac{\Delta Q}{\Delta t}$$

The rate of change of entropy is then

$$\frac{\Delta S}{\Delta t} = \frac{\Delta Q / \Delta t}{T} = \frac{12 \times 10^3 \text{ W}}{20 + 273 \text{ K}} = \underline{41 \text{ W/K}}$$

21-65. The potential energy of the water is dissipated as heat at a rate

$$\frac{\Delta m}{\Delta t} gy = \frac{\Delta Q}{\Delta t}$$

The rate of change of entropy is then

$$\frac{\Delta S}{\Delta t} = \frac{\Delta Q / \Delta t}{T} = \frac{\left(\frac{\Delta m}{\Delta t}\right) gy}{T} = \frac{\left(5700 \text{ m}^3/\text{s}\right)\frac{1000 \text{ kg}}{1 \text{ m}^3}\left(9.81 \text{ m/s}^2\right)(50 \text{ m})}{20 + 273 \text{ K}} = \underline{9.5 \times 10^6 \text{ W/K}}$$

21-69. In the free expansion of the gas, we remember from the First Law of Thermodynamics that

$$\Delta E = \Delta Q - W = \Delta Q - p\Delta V$$

Since the process is an isothermal one, $\Delta E = 0$ J and $\Delta Q = p\Delta V$

If we consider the system by adding and removing infinitesimally small amounts of heat, the change in entropy is

$$dS = \frac{dQ}{T} = \frac{pdV}{T} = nR\frac{dV}{V}$$

$$\Delta S = \int dS = \int_V^{V'} nR\frac{dV}{V} = nR\ln\frac{V'}{V} = (1.00 \text{ mole})(8.31 \text{ J/(mole} \cdot \text{K)})\ln\frac{2.0 \text{ liters}}{1.0 \text{ liter}} = \underline{5.8 \text{ J/K}}$$

21-73. Because we are mixing equal amounts of warm and cool water, the final temperature will be 50°C, the mid-point between the two temperatures. The total entropy change is $\Delta S = \Delta S_1 + \Delta S_2$ where ΔS_1 is the change in entropy of the initially cooler water and ΔS_2 is the change in entropy of the initially warmer water.

$$\Delta S_1 = mc\ln\frac{T_2}{T_1} = (1.0 \text{ liter})\left(\tfrac{1 \text{ kg}}{1 \text{ liter}}\right)(4187 \text{ J/(kg} \cdot \text{°C)})\ln\left(\frac{50 + 273 \text{ K}}{20 + 273 \text{ K}}\right) = 408 \text{ J/K}$$

$$\Delta S_2 = mc\ln\frac{T_2}{T_1} = (1.0 \text{ liter})\left(\tfrac{1 \text{ kg}}{1 \text{ liter}}\right)(4187 \text{ J/(kg} \cdot \text{°C)})\ln\left(\frac{50 + 273 \text{ K}}{80 + 273 \text{ K}}\right) = -371 \text{ J/K}$$

$$\Delta S = \Delta S_1 + \Delta S_2 = 408 \text{ J/K} - 371 \text{ J/K} = \underline{37 \text{ J/K}}$$

21-77. For n moles of a diatomic gas at temperature T, the internal energy is $E = \frac{5}{2}nRT$

The change in internal energy is $\Delta E = \Delta Q - W = -W$

Because work is done on the system, the sign for the work is negative, $W = -2500$ J

Combining the above equations, gives

$$\Delta E = \frac{5}{2}nR\Delta T = -W$$

$$\Delta T = \frac{-2W}{5nR} = \frac{-2(-2500 \text{ J})}{5(1 \text{ mole})(8.31 \text{ J/(mole} \cdot \text{K)})} = \underline{120 \text{ K}}$$

21-85. (a) Beginning with the point at the upper left, the gas undergoes an isobaric expansion in step 1 as the volume is increased, followed by a isovolumetric reduction of pressure in step 2 as the temperature is reduced. The gas is then compressed isobarically in step 3 by reducing the volume, before an isovolumetric increase in pressure in step 4 by increasing the temperature.

(b) Step 1: $W_1 = p\Delta V = (1.5 \times 10^5 \text{ N/m}^2)(0.021 - 0.007 \text{ m}^3) = \underline{2100 \text{ J}}$

Step 2: $W_2 = \underline{0 \text{ J}}$ (no work is done in an isovolumetric process)

Step 3: $W_3 = p\Delta V = (0.5 \times 10^5 \text{ N/m}^2)(0.007 - 0.021 \text{ m}^3) = \underline{-700 \text{ J}}$

Step 4: $W_4 = \underline{0 \text{ J}}$ (no work is done in an isovolumetric process)

(c) Step 1: At the beginning of step 1, the temperature is

$$P_1 V_1 = nRT_1$$

$$T_1 = \frac{P_1 V_1}{nR} = \frac{(1.5 \times 10^5 \text{ N/m}^2)(0.007 \text{ m}^3)}{(1.00 \text{ mole})(8.31 \text{ J/(mole} \cdot \text{K)})} = 126 \text{ K}$$

At the end of step 1, the temperature is

$$T_2 = \frac{P_2 V_2}{nR} = \frac{(1.5 \times 10^5 \text{ N/m}^2)(0.021 \text{ m}^3)}{(1.00 \text{ mole})(8.31 \text{ J/(mole} \cdot \text{K)})} = 379 \text{ K}$$

$Q_1 = nC_p\Delta T_1 = (1 \text{ mole})(20.8 \text{ J/(mole} \cdot \text{K)})(379 \text{ K} - 126 \text{ K}) = \underline{5260 \text{ J}}$

Step 2: At the end of step 2, the temperature is

$$T_3 = \frac{P_3 V_3}{nR} = \frac{(0.5 \times 10^5 \text{ N/m}^2)(0.021 \text{ m}^3)}{(1.00 \text{ mole})(8.31 \text{ J/(mole} \cdot \text{K)})} = 126 \text{ K}$$

$Q_2 = nC_V\Delta T_2 = (1 \text{ mole})(12.5 \text{ J/(mole} \cdot \text{K)})(126 \text{ K} - 379 \text{ K}) = \underline{-3160 \text{ J}}$

Step 3: At the end of step 3, the temperature is

$$T_4 = \frac{P_4 V_4}{nR} = \frac{(0.5 \times 10^5 \text{ N/m}^2)(0.007 \text{ m}^3)}{(1.00 \text{ mole})(8.31 \text{ J/(mole} \cdot \text{K)})} = 42 \text{ K}$$

$Q_3 = nC_p\Delta T_3 = (1 \text{ mole})(20.8 \text{ J/(mole} \cdot \text{K)})(42 \text{ K} - 126 \text{ K}) = \underline{-1750 \text{ J}}$

Step 4: $Q_4 = nC_V\Delta T_4 = (1 \text{ mole})(12.5 \text{ J/(mole} \cdot \text{K)})(126 \text{ K} - 42 \text{ K}) = \underline{1050 \text{ J}}$

In part (b), the net work done is 2100 – 700 J = 1400 J. From part (c), the sum of the four quantities of heat also equals 1400 J.

(d) The efficiency of this four-step cycle is

$$e = \frac{W}{Q_2} = \frac{1400 \text{ J}}{3160 \text{ J}} = \underline{0.44 \text{ or } 44\%}$$

21-89. The boiling point temperatures and heats of vaporization are given in Table 20.4.

$$\Delta S_N = \frac{\Delta Q}{T_{boil}} = \frac{mL_V}{T_{boil}} = \frac{(1.00 \text{ kg})(2.00 \times 10^5 \text{ J/kg})}{-196 + 273 \text{ K}} = \underline{2600 \text{ J/K}}$$

$$\Delta S_O = \frac{\Delta Q}{T_{boil}} = \frac{mL_V}{T_{boil}} = \frac{(1.00 \text{ kg})(2.1 \times 10^5 \text{ J/kg})}{-183 + 273 \text{ K}} = \underline{2300 \text{ J/K}}$$

$$\Delta S_H = \frac{\Delta Q}{T_{boil}} = \frac{mL_V}{T_{boil}} = \frac{(1.00 \text{ kg})(4.5 \times 10^5 \text{ J/kg})}{-253 + 273 \text{ K}} = \underline{2.25 \times 10^4 \text{ J/K}}$$

Hydrogen is largest and oxygen the smallest.